OUR BETTER NATURE

"At a time when many key habitats and species are declining in Vermont and throughout the world, *Our Better Nature* serves as a well-crafted and timely reminder of the critical importance of redoubling efforts to protect and enhance working and wild landscapes. A must-read for anyone interested in learning more about the nature of Vermont and many of the efforts to protect it for future generations."

— **ANTHONY D'AMATO, PROFESSOR, DIRECTOR OF FORESTRY PROGRAM, UNIVERSITY OF VERMONT**

"Faced by crises in climate, biodiversity and human well-being, we need a new mode of living that secures the Earth and all life intact. Vermont has responded by seizing on the natural rewilding of the past century, advancing a Conservation Design vision for the state's future, and embracing the lives of so many individuals who are forging that future throughout this resilient landscape. Through diverse voices, *Our Better Nature* shares this grand, humbling, and hope-filled story."

— **DAVID FOSTER, HARVARD FOREST, COORDINATOR OF THE WILDLANDS & WOODLANDS INITIATIVE**

"What a rare and lovely treat, to get to read such interesting and thoughtful essays written about nature, by people who have dedicated their lives to living close to and understanding the land of green hills and silver waters of Vermont. Through the eyes of these great writers and thinkers (and doers!), we get a birds-eye view of the beautiful and complex quilt of life in the Green Mountain state. Providing lessons that apply far beyond Vermont's borders, these writers give us both a clear warning that this quilt is unraveling and invite us to join in celebrating, and saving, our natural world."

— **DAVID K. MEARS, EXECUTIVE DIRECTOR, AUDUBON VERMONT**

OUR BETTER NATURE

Hopeful Excursions in Saving Biodiversity

Edited by Curt Lindberg and Eric Hagen

©2022 Curt Lindberg and Eric Hagen
All rights reserved

Published by Vermont Alliance for Half-Earth, Northeast Wilderness Trust, Vermont Natural Resources Council, and the Lintilhac Foundation

vermontallianceforhalfearth@gmail.com

ISBN: 979-8-9855315-0-3

Book design by Kevin Cross
Cover photos by Sean Beckett

Printed in Vermont by Villanti Printers using VOC-free, soy-based inks; 100% certified renewable energy; and paper certified by the Forest Stewardship Council®

To the lands, waters, plants, and other creatures of this Earth, and to all of its caretakers—past, present, and future.

*And to Edward O. Wilson
(June 10, 1929–December 26, 2021)
for inspiring us to become better caretakers.*

OUR BETTER NATURE

Foreword
George Schenk — ix

Introduction
Eric Hagen and Curt Lindberg — xv

PART ONE ~ LESSONS IN BIODIVERSITY

Nature Needs More: Insight and Inspiration from E. O. Wilson
Curt Lindberg, with Maddie Lindberg — 3

A Rewilding Story
Tom Butler — 17

Conservation in Vermont: Wildness to Devastation, Opportunity to Intention
Elizabeth Thompson — 33

Adventures in Forest Carbon: Conservation, Carbon Capture, and Our Climate Future
Annie Faulkner — 49

High Flyers
Prucia Buscell and Joe Roman — 67

Saving the Forest by Learning the Trees
Alicia Daniel — 85

Half-Earth Vermont: An Origin Story
Steven Shepard — 101

PART TWO ~ A SHARED LIFE

Guests at Nature's Table: Food for People and a Home for Wildlife
Eric Hagen   117

Roots in the Land: Generations of Mentorship
Eric Hagen   123

A Field That Bloomed: Relearning What's Been Lost
Eric Hagen   129

The Winding Path of Reconciliation
Andlea Brett   139

An Unusual Forest: You Can Manage Land and Love It Too
Eric Hagen   143

Forever Wild: Quiet Refuge and Space to Listen
Eric Hagen   153

Rewilding Firefly Hill: Biodiversity Conservation in a Backyard
Eric Hagen   163

PART THREE ~ TAKING ACTION

A Vermonter's Guide for Protecting Biodiversity:
At Home, in Your Town, Across the State
Eric Hagen   173

Building Movements to Protect Biodiversity:
The Diffusible, the Stewardable, and the Possible
Arvind Singhal and Eva Duedahl   193

Afterword
Doug Tallamy   207

Acknowledgments   213

About the Contributors   215

. . . all these unlovable species with funny names are connected to a system, and if the system is in trouble, so are we.

MOLLIE BEATTIE

Foreword

GEORGE SCHENK

Nature is in trouble. Almost everywhere biologists have looked they have found the biosphere in decline. The numbers are staggering:

Between 1970 and 2016, populations of mammals, birds, reptiles, amphibians, and fish declined an average of 68 percent;[1] In the last 50 years North America has lost 3 billion songbirds;[2] 90 percent of the world's marine fisheries that for millennia have been an important source of nutritious and delicious food for the human family have been either overfished or are fully exploited;[3] and a broad range of insect species has declined by 40 percent.[4] Forests, so essential for wildlife habitat and the regulation of the atmosphere, have been and continue to be cut, burned, and fragmented to the detriment of the native species that depend on them,[5] and 98 percent of the native vegetation of North America's largest biome, the prairie, has been overturned by modern agriculture that has replaced a complex self-regenerative grassland ecosystem with monoculture cropping dependent on synthetic fertilizers and pesticides.[6]

The degradation of the biosphere began slowly at first with localized changes to the environment for human benefit. Natural landscapes became urban settlements; riparian lands were drained and plowed; estuaries became harbors; forests were felled for wood, crops, and pasture; and rivers, lakes, the oceans, and the atmosphere were used as convenient dumping grounds for wastes. It became a common view that all of nature was a resource for human enrichment and comfort, a resource we were entitled to and held absolute sovereignty over. This view, coupled with advances in technology and an ever-burgeoning human population, has precipitated a world-wide crisis of the biosphere that is putting at risk the essential abundance, beauty, and creativity of life on Earth.[7]

Evolution is a dynamic continuum in which species emerge and disappear. It has been estimated that since the Cambrian diversification of life some 600 million years ago, 99 percent of species have gone extinct.[8] For most of that time the average extinction rate ebbed and flowed modestly. However, there have been five cataclysmic extinction events, the last happening 65–66 million years ago when a 10-mile-wide asteroid slammed into the coastal waters of present-day Mexico. The impact threw up a cloud of dust that chilled the planet and triggered a period of tidal waves, earthquakes, and volcanism that killed off 70 percent of the existing species, including all the remaining Mesozoic Era dinosaurs. It took about 10 million years for biodiversity to recover from this calamity.[9] The current extinction rate is estimated to be a thousand times greater than the average or normal background rate,[10] which has suggested to some that we have recently entered into a sixth extinction event, this time caused by human-driven changes to the environment.

When species become rare or go locally extinct ecosystem function is degraded to the detriment of all. This is because species do not exist in isolation. They are always part of larger arrays of interacting matrices of life ecologists call "ecosystems." Healthy ecosystems provide a dynamic framework of life that benefits all living members of the system—including us *Homo sapiens*.

As for how or why the biosphere is being degraded . . .

THE SCIENTIFIC ANSWER is summarized by the acronym HIPPO, first coined by E. O. Wilson:

Habitat destruction and fragmentation. Urban/suburban development is certainly an important factor, but the leading contributor is agriculture, which world-wide uses 50 percent of the ice-free land base.[11]

Invasive species. The international commerce of plants, animals, and their products has greatly accelerated the spread of non-native species which can become disruptive, in some cases profoundly so, to local ecosystems by crowding out native species that are evolutionarily keyed to other local species.

Pollution. There are a lot of pollutants out there but perhaps none as important as the climate-changing gases carbon dioxide and methane.

Population (human). The current human population is estimated to be two to four times the ecologically sustainable carrying capacity of the Earth.[12] So often the ecological benefits of advances in the efficiency or sustainability of a product or service are overwhelmed by the increasing number of people who want that product or service.

Over-harvesting by hunting and fishing. The bushmeat trade and poaching are decimating wildlife populations, especially in tropical forests, and commercial fishing fleets are virtually sweeping the oceans clean of many species of fish, changing marine ecosystems in fundamental ways.[13]

These five factors often act synergistically with one another compounding or magnifying their destructive effect on the biosphere.

THE PERSONAL ANSWER is . . . *me*. I am not alone in this regard, but my life and livelihood are part of the story.

My truck, made of energy-intensive metals and plastics fueled by long-buried carbon, so easy to use—just hop right in and off I go without a thought or a care; the clothes upon my back, made halfway around the world because it's cheaper that way (cheaper for me but more costly to nature).

My comfortable home, defiant of the natural rhythms of the world—warm and snug in winter, light-filled at night.

My food, grown on a far-away field, plowed and sprayed to make it good for one or two crops but poor for almost every other living thing.

My jet plane vacations—wonderful and culturally enriching to me, so costly to the world I earnestly seek to see and know.

My livelihood, which depends on employees who drive to work and customers who can really only come to me by car; the trucked-in supplies, the trucked-out trash and recycling (how much really gets recycled?); the lights and heat and refrigeration, the paint, paper, and plastics . . . the water, oh my, the great quantities of water, pumped by an electric motor from an ancient aquifer to quench my thirst . . . and wash away the dirt of my life.

Almost everything I do . . . or have . . . or want, is in opposition to nature. My modern economy is in direct conflict with my essential ecology.

SO, WHAT TO DO? While thinking about all of this out in the garden this Spring, my mind went to a very dark place. I wrote in my journal: "I am the enemy of every green and living thing." I am a thief of everything I purport to love, not by intent perhaps, but by habit and by my way of making a living. I was not born in opposition to nature. My habits and patterns of living are inculturated, a learned worldview that sees nature without intrinsic or sovereign value unto itself, that its only value is its utility to me. This is an amoral relationship with nature. In the end, the care of nature is a moral choice.

Survivors of armed conflict often face moral choices. The Indian war of independence pitted Hindus and Muslims against one another. Atrocities were committed on both sides. Many children were orphaned. In a story attributed to Gandhi, after the war a Hindu man came to Gandhi in despair: "I will certainly go to Hell," he said. "I killed a Muslim child." Gandhi felt the man's deep remorse and replied, "I know a way out of Hell. Go and adopt a Muslim orphan and raise him as a Muslim."

My life and livelihood, shaped by my sense of entitlement and supremacy, have made orphans of nature. My work now is to make a better, safer place for the surviving children of nature—to nurture them by their own creed in their own way, in their own time—and to give voice to their great worth in the halls of humanity. I am sure I will be imperfect in this effort but that is no reason not to try.

E. O. Wilson wrote that the degradation of the biosphere is the greatest threat to humanity's future. The idea of putting aside half of the Earth for nature, as Wilson has calculated is necessary to maintain a healthy diversity of life, can seem overwhelming for any individual. A few years ago I met Wilson, and when I asked him about my work to increase the biodiversity in my little garden—the compost and mulch that feed and shelter the soil biota, the intentionally un-mown grass and piles of brush and stones that make such good refugia for all manner of little lives, the flower beds planted to benefit bees and butterflies and other helpful insects, the standing posts that make good vantage points for songbirds on the hunt, the light touch of hand cultivation, and the pathways for children and their parents to explore and be immersed in the abundance and beauty of nature—he said that everything that enriches life helps.

Humanity is biological. Our health and happiness are intrinsically linked to the health and well-being of the biosphere. Today everywhere we look we find the biosphere in trouble because of human-created changes to the environment. The trouble is real, it is now, it is here and everywhere, and the trouble is us.

Business as usual is not working for nature. Nature immediately needs our creative participation. It needs the big help of governments, institutions, and private enterprise—help we can advocate for with our voice, vote, and buying power—and it needs the backyard help of individuals making a place for the wild where, in Doug Tallamy's words, we "live, work and farm."[14] Some of this work will be easy—leaving autumn leaves around the base of trees and shrubs (habitat that helps many insects survive the winter)—and some of it will be hard—challenging cherished expectations and assumptions of an ever-expanding economy and the idea that we are entitled to impose our will on the world without regard to what it means to other living things.

There is still much to be learned about the ways of nature but we know enough to start rethinking our relationship with the other lives-of-the-world. This important and wonderful book can help us to a more ecological view of our own lives, and a better understanding of the great worth of all green and living things.

REFERENCES

1. WWF. (2020). *Living planet report 2020: Bending the curve of biodiversity loss*. WWF.
2. Rosenberg, K. V., Dokter, A. M., Blancher, P. J., Sauer, J. R., Smith, A. C., Smith, P. A., et al. (2019). Decline of the North American avifauna. *Science*, 366(6461).
3. WWF. (2015). *Living blue planet report: Species, habitats and human well-being*. WWF.
4. Sánchez-Bayo, F., & Wyckhuys, K. (2019, January). Worldwide decline of the entomofauna: A review of its drivers. *Biological Conservation*.
5. Rogan, J. E., & Lacher, T. E. (2018). Impacts of habitat loss and fragmentation on terrestrial biodiversity. *Reference module in earth sciences and environmental systems*. Elsevier.
6. Manning, R. (2004). *Against the grain: How agriculture has highjacked civilization*. North Point Press.
7. Crist, E. (2019). *Abundant earth: Toward an ecological civilization*. University of Chicago Press.
8. Wilson, E. O. (2016). *Half-Earth: Our planet's fight for life*. Liveright Publishing Corporation.
9. *Ibid.*
10. *Ibid.*
11. Ellis, E. C., Goldewijk, K. K., Siebert, S., Lightman, D., & Ramankutty, N. (2010). Anthropogenic transformation of the biomes, 1700 to 2000. *Global Ecology and Biogeography*, 19(5).
12. Crist, *Abundant earth*.
13. Crist, *Abundant earth*.
14. Tallamy, D. W. (2019). *Nature's best hope: A new approach to conservation that starts in your yard*. Timber Press, p. 36.

The forest itself is part of much larger cycles, the building of soil and the migration of species and the circulation of oceans. The source of clean air and pure water and good food. There is a necessary wisdom in the give and take of nature—its quiet agreements and search for balance. There is an extraordinary generosity.

SUZANNE SIMARD

Introduction

ERIC HAGEN AND CURT LINDBERG

The abundance and diversity of life on Earth is in a state of alarming decline. Since the beginning of human civilization the amount of life on Earth, as measured by weight, has fallen by half.[1] Not only has the total amount of life on Earth fallen, but the distribution has shifted drastically. While the pattern is not uniform across the globe, humans and our livestock now make up 96 percent of all mammalian biomass on the planet, while wild mammals account for only 4 percent.[2] Undoubtedly the most up-to-date and authoritative source of information on the biodiversity crisis comes from the *2019 Global Assessment Report* by the Intergovernmental Science-Policy Platform on Biodiversity and Ecosystem Services (IPBES).[3] This report was led by over 150 experts and drew upon the help of 350 contributing authors. It reviewed over 15,000 scientific papers, and was accepted by more than 130 governments. Two of the most powerful findings are that an estimated 1 million species currently face global extinction, and nothing short of "transformative changes across economic, social, political and technological factors" can avert catastrophic global ecosystem collapse.[4]

In the face of such daunting challenges we offer a message of hope from the state of Vermont, in the northeastern United States. The book you are about to read was created by the Vermont Alliance for Half-Earth, a group of Vermont citizens inspired by acclaimed biologist E. O. Wilson. Wilson's message is that in order to halt the decline of nature and the extinction of species we need to dedicate half of the Earth's surface to nature. We decided to take that enormous goal and apply it to our own home region. What we've found is that it is absolutely possible to protect biodiversity in Vermont, though the best methods for getting there aren't exactly what we were expecting (hint: we don't have to cut Vermont, or the Earth, in half). We have also found that there are already many amazing people, organizations, and state-led programs dedicated to protecting biodiversity in Vermont. So, we decided to share their stories with new audiences in order to accelerate biodiversity protection to the level that is needed. Our aim with this book is to share some of what we've learned in Vermont and inspire more action to protect biodiversity here at home and around the world.

As the 2019 IPBES report indicates, there are many complex interactions between economic, social, political, and technological factors that all need to change dramatically for people and the rest of nature to cohabitate this Earth sustainably. National and international patterns of trade, regulations, sovereign and ecological debts, austerity, perverse financial incentives, extraction, consumption, inequality, and issues of land tenure all need to be addressed in order for biodiversity loss to be halted and the implementation of conservation science to be effective and equitable. There are so many direct and indirect drivers of biodiversity loss that need to be addressed, but we don't discuss all of them in this book. Instead, we focus here on the well-designed conservation science and educational efforts found in Vermont that are preparing us to effectively protect biodiversity through the uncertainties of the coming decades and beyond. We emphasize the need for nature conservation initiatives everywhere that create more opportunities for people to be in closer relation to nature, that respect indigenous land rights and responsibilities, that are based in community decision making, and that rely on the best conservation science for guidance.

This book is organized into three parts. The first is a collection of seven essays by members and friends of the Vermont Alliance for Half-Earth on their own topics of expertise within the movement to protect and care for biodiversity. Here we look at the role of E. O. Wilson in inspiring our organization; the power of a global story of rewilding; the history and guiding science of conservation in Vermont; the dual role of the carbon cycle in biodiversity conservation and climate change mitigation; successful models of educating students and professional naturalists; and our organization's origin and guiding principles.

The second part of the book is a collection of six stories about people in Vermont and their connections with our landscape. These stories come out of the 2019/2020 graduate work of one of us, Eric Hagen, who collaborated with the Vermont Alliance for Half-Earth to create educational material to explain the science of biodiversity conservation through the personal stories of people who live in Vermont. Each story follows a person or a couple on a walk around the land they know best, and features photographs of the places and creatures they talk about, as well as core ideas from conservation science and land stewardship. The most obvious reason to share these is that stories are one of the best ways to convey information because they enliven abstract concepts and engage the heart and imagination together with the mind. A second theme that we want to convey through these stories is that meaningful relationships with nature can contribute to the well-being of people and nature alike, as well as sustain the action and attitudes needed to live sustainably with the rest of life.

The third part of the book is action-oriented, and features a how-to guide for protecting biodiversity before ending with an essay on movement building by two experts on how ideas spread through social networks. The guide for protecting biodiversity was originally written to give clear instructions to Vermonters for protecting, enhancing, and restoring biodiversity in our yards, in our towns, and across the state. Though written for Vermont,

the guide provides ideas for protecting biodiversity anywhere, and our hope is that readers outside of Vermont can take the strategies outlined in this guide and shape them to fit the places and communities where they live. The final essay on movement building was written to inspire readers to think beyond the individual actions they can take on their land or in their locality, and consider what else they can do to reach others, and so contribute to the social movement to restore nature.

Each of the three sections build upon each other and offer unique insights into the effort to accelerate the protection and care of biodiversity, but they all can also be read as stand-alone pieces. We hope that you find in these pages the inspiration and knowledge to lend your hands to the effort to care for the natural world wherever you are, and to share what you learn with others. By tending to our own better nature, and dedicating ourselves to meaningful relationships and to effective action, we will be able to live well with the rest of life on this one precious planet.

REFERENCES

1. Bar-On, Y. M., Phillips, R., & Milo, R. (2018). The biomass distribution on Earth. *Proceedings of the National Academy of Sciences*, 115(25).

2. *Ibid.*

3. IPBES. (2019). *Global assessment report on biodiversity and ecosystem services of the Intergovernmental Science-Policy Platform on Biodiversity and Ecosystem Services.* IPBES secretariat.

4. *Ibid.*

*People need wild places . . . to be surrounded
by a singing, mating, howling commotion
of other species, all of which love their
lives as much as we do ours.*

BARBARA KINGSOLVER

PART ONE

Lessons in Biodiversity

A collection of essays from local experts and members of the Vermont Alliance for Half-Earth. Each essay focuses on a different critical aspect of our efforts to protect biodiversity.

There can be no purpose more inspiring than to begin the age of restoration, reweaving the wondrous diversity of life that still surrounds us.

EDWARD O. WILSON

Nature Needs More

Insight and Inspiration from E. O. Wilson

CURT LINDBERG, WITH MADDIE LINDBERG

At the conclusion of a talk on biodiversity in a conference room packed with healthcare professionals, Edward O. Wilson noticed something. A slight, older woman in the back row slowly raised her hand (it barely got above her shoulder) to pose a question to this world-renowned scientist. Spotting this tentative signal, Dr. Wilson stepped off the stage and walked to the back corner of the room to have a private talk with her. It was a small act of compassion. And not the only one that day.

I (Curt) can relate. It took me some time to muster the nerve to place my first call to Dr. Wilson. When I did dial his Harvard number I figured I'd have to navigate through a phalanx of secretaries and graduate assistants. You can imagine my surprise when I heard, "Hi, this is Ed." It then took me what seemed like an eternity to gather my wits and begin the conversation.

Before the conference I asked Ed if he'd be willing to stay after the session and sign copies of his book *The Diversity of Life*. We had purchased copies for the attendees—all 350 of them. "Why don't I come early and sign them all?" he responded. (If you were fortunate, an autographed copy also included a little hand-drawn ant.)

While certainly not a major event in Dr. Wilson's career, his presentation had a big impact on those who attended. They learned about important new insights into healthy ecosystems and about the man himself, the qualities of this esteemed scientist on display throughout the morning.

The little things

They saw an astute observer of life, especially the little things.
The raised hand, barely visible in the back of the room.

At the age of seven, the young Wilson spent the summer in a shorefront house in Paradise Beach, Florida. As he explains in his autobiography *Naturalist*, all day, every day he was out searching for life in the ocean and seashore, loving the adventure. An accident while fishing set him on a path to become a certain kind of naturalist. He was fishing for pinfish and yanking them out of the water when they struck the bait. One pull was too hard and a pinfish flew through the air and a spine on its dorsal fin penetrated the pupil of his right eye. Wilson ignored the painful injury. Months later, when his parents noticed that his pupil had begun to cloud over, they took him to an ophthalmologist who removed the damaged lens. This left him with full sight in only one eye. Fortuitously, eyesight in his left eye proved to be especially acute at close range, 20/10. With sight in only one eye Wilson also lost depth perception, which he said made him a wretched birdwatcher. With the benefit of acute close-up vision in the left eye and the loss of stereoscopy, the options for his scientific focus became clear. "The attention of my surviving eye turned to the ground. I would thereafter celebrate the little things of the world, the animals that can be picked up between thumb and forefinger and brought close for inspection."[1] An accidental encounter with a pinfish led to a career as an entomologist. He would become the world's leading authority on ants, credited with discovering 450 new ant species. Yet, his scientific gaze was not restricted to little things.

Beyond little

They saw a scientist deeply curious about all things living.
He seemed interested in what the woman in the back of the room wanted to talk about.

Wilson's ever-youthful curiosity led to discoveries both big and small, from new ant species to new scientific theories and disciplines, in both the natural and social sciences.

His investigations of ant behavior led to groundbreaking insights into self-organization, a key concept in the science of complexity. He showed how simple interactions in a system (ants, using pheromones, convey basic information: follow this path to food, watch out, danger ahead)—and among the elements of a system, lead at higher levels in the system to emergent properties not explainable by behaviors of the system's separate components. While individual ants are capable of only a limited range of simple functions, ant colonies, called superorganisms by Wilson, exhibit incredibly complex, adaptable behavior. So successful are ants and other social insects that by impact and biomass they have become "dominant elements in land ecosystems."[2]

Studies of the social behavior of certain insect species stimulated Wilson to formulate an entirely new scientific discipline, sociobiology, the study of the biological basis for social behavior in all kinds of organisms, including *Homo sapiens*. His book, *Sociobiology: The New Synthesis*, was the foundational text in this new field and a basis for the development of the related new field of evolutionary psychology.

His early research had profound implications for conservation. In 1967, in collaboration with Robert MacArthur, he coauthored *The Theory of Island Biogeography*. In it, they demonstrated that island size (and think of nature reserves as islands on land) and distance from the mainland correspond with the number of species and the populations of species islands can support.[3] Over time, conservationists came to appreciate the implications of

E. O. Wilson explores the return of life in Gorongosa National Park, Mozambique.

this research. Carving ecosystems into smaller, isolated fragments simultaneously fragments large populations of species and isolates them, making them vulnerable to local extinction from environmental stress that larger populations could withstand. These findings have become cornerstones of conservation biology. Today, conservationists talk of reducing ecosystem fragmentation, knitting together larger tracts of conserved land, and setting aside wildlife corridors.

They saw a leader in biodiversity scholarship. His remarks led attendees to think more expansively about the bases for ecosystem and human health.

When *The Theory of Island Biogeography* came out, the concept of biodiversity was in no one's vocabulary, but this work helped put science, and scientists like Wilson, on the road to its coinage. An early, and influential, contribution from Wilson was his 1985 article "The Biological Diversity Crisis: A Challenge to Science." Highlighted in this landmark paper were important insights paired with calls to action.[4] We really know little about life on Earth. Only 1.7 million of the estimated 10 million species on Earth have been identified and formally named. Wilson's call to action: create a catalog of all life (later in his career he was instrumental in the development

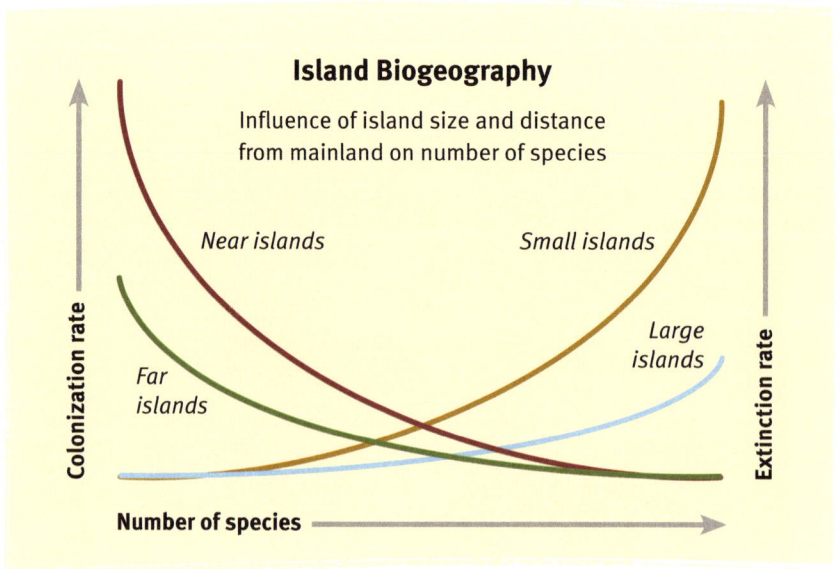

of the Encyclopedia of Life, an online resource with a webpage for every known species). The crisis noted in the title is the human-caused and accelerating loss of biological diversity from environmental degradation and habitat destruction, causing a rate of species extinction about four hundred times greater than in recent geologic time. His call to action: an international drive to save biological diversity.

With awareness of biodiversity growing in scientific circles the National Academies of Science and the Smithsonian Institution sponsored a National Forum on BioDiversity in 1986. Wilson gave one of the keynote addresses and later served as the editor of conference proceedings. *Biodiversity* was published in 1988 and soon became a best seller, contributing to growing concern around the world and introducing the term into the conservation lexicon.[5] This concern spurred action. At the 1992 Earth Summit in Rio de Janeiro 150 governments adopted the Convention on Biological Diversity and its call for the conservation of biological diversity, the sustainable use of its components, and the fair and equitable sharing of the benefits from the use of genetic resources. In 1992, Wilson's book *The Diversity of Life* was published. In it he defined biodiversity as "the variety of organisms considered at all levels, from genetic variants belonging to the same species through arrays of species to arrays of genera, families, and still higher taxonomic levels; includes the variety of ecosystems, which comprise both the communities of organisms within particular habitats and the physical conditions under which they live."[6]

To half-earth

In *The Future of Life,* published in 2002, he went beyond his general calls to halt species extinction by broaching the idea of setting aside half of the planet for nature. He wrote, "At the risk of being called an extremist, which on this topic I freely admit I am, let me suggest 50 percent. Half the world for humanity, half for the rest of life, to create a planet that is both self-sustaining and pleasant."[7]

Fourteen years later, in 2016, and 49 years after the publication of *The Theory of Island Biogeography*, Wilson wrote *Half-Earth: Our Planet's Fight for Survival*. It is his most fervent and fully formulated call to stave off the mass extinction of species and preserve Earth's diminishing biodiversity. He contends that as nature reserves on land and in the sea grow in size or are created by connecting smaller plots, they can harbor more ecosystems and the species they comprise. Calculations by biologists and mathematicians show that more than 80 percent of existing species can be protected by dedicating half of the planet to nature (Wilson later wrote that increasing refuges for nature from what exists now—15 percent of the land and 3 percent of the sea—to half could save upward of 90 percent of existing species.) Intelligently located, such an allocation could also support a comprehensive representation of Earth's ecosystems and the species that inhabit them.

In his half-earth appeal Wilson emphasizes that biodiversity "forms a shield protecting each of the species that together compose it."[8] Because of intricate interdependencies in ecosystems, this shield can be crippled. As the number and types of species go near or to extinction, the species whose lives depend on them are also threatened, setting in motion an extinction cascade. An example Wilson uses is the cascade of loss triggered by the demise of the American chestnut. The caterpillars of seven moth species depended solely on the leaves of the American chestnut for food. As logging and an Asian fungal blight ravaged this once prevalent tree species, these moths were doomed. Also impacted by the demise of the American chestnut was the passenger pigeon. These birds depended on the yearly crop of chestnuts to feed their vast flocks. The weakened species then succumbed to another fatal pressure: unrelenting human hunting. Once the most abundant bird in North America and possibly the world, numbering in the billions, the species was lost in 1914 when the last known passenger pigeon died in a Cincinnati zoo.

Scientists know that at some point as more and more species disappear, entire ecosystems will collapse. We are moving inevitably closer to such tipping-point catastrophes as the human-caused species extinction rates rise to what scientists now estimate to be a thousand times greater than the pre-human level. Wilson feared that at this rate of destruction 50 percent of Earth's remaining species will become extinct or be on the edge of extinction by the end of the century. And these lost species will be lost forever.

On to activism

They saw a scientist who added activism to his professional resume and then embraced this new role. At the conference he made a convincing link between human and ecosystem health and appealed to the healthcare professionals to advocate for both.

Many scientists struggle to break from the tradition of the scientist as the neutral, detached observer, fearing it will tarnish perspectives and professional reputations. Wilson struggled with this too.

Wilson attributed his debut as an activist to a question posed to him and six other Harvard University professors in 1980 by the editors of *Harvard Magazine*: *What do you believe will be the most important problem facing the world in the coming decade?* Poverty was the most frequent response. Other nominations included the nuclear threat and the welfare state in the United States. Wilson went in another direction.

> The worst thing that can happen, *will* happen, is not energy depletion, economic collapse, limited nuclear war, or conquest by a totalitarian government. As terrible as these catastrophes would be for us they can be repaired within a few generations. The one process ongoing in the 1980s that will take millions of years to correct is the loss of genetic and species diversity by the destruction of natural habitats. This is the folly our descendants are least likely to forgive us.[9]

Reflecting later in life on his adoption of this new mantle as activist, Wilson confided he was "unforgivably late in arriving."[10]

Like all his other serious endeavors, he proceeded vigorously.

Wilson joined the boards of World Wildlife Fund-U.S. and Conservation International, he lectured, he advised American presidents, he advocated for good conservation policy and, of course, he wrote.

> *They heard an accomplished writer, able to reach a wide variety of audiences. He came early to autograph books for those in the audience. Like* Diversity of Life, *the book he autographed, more of his books sought to reach and influence a public audience, including children.*

How many Pulitzer Prize winners can you name who won this award for a scientific text? A book Wilson wrote with Bert Hölldobler, *The Ants*, is both the definitive work on the planet's most significant insect family and the only academic scientific text to win this prize. *On Human Nature* also garnered a Pulitzer Prize. Among the books aimed at the public and penned to spur activism and action are: *The Future of Life*; *The Creation: An Appeal to Save Life on Earth*; *A Window on Eternity: A Biologist's Walk Through Gorongosa National Park, Mozambique*; and, of course, *Half-Earth: Our Planet's Fight for Life*.

These and other publications have been enormously influential, likely because of Wilson's rare combination of skills—those of a revolutionary investigator of life, of a discipline-crossing activist, and of an artistic wordsmith.

Linger with his words to Henry David Thoreau from *The Future of Life*.

> You searched for essence at Walden and, whether successful in your own mind or not, you hit upon an ethic with a solid feel to it: nature is ours to explore forever; it is a crucible and refuge; it is our natural home; it is all

Like Wilson, Sean Beckett has an eye for small creatures, human and avian. Sean, a naturalist at North Branch Nature Center, Montpelier, Vermont, was leading a group to a wooded area where mist nets had been arrayed to capture northern saw-whet owls on their fall migration south. He noticed a small young girl close on his heels. When the group returned to the conference room where the owls would be examined, measured, and banded (North Branch participates in a northern saw-whet owl research project), Sean asked Maddie, my eight-year-old granddaughter, to come up front and be his assistant. She helped examine one of the owls. Maddie wrote down the results—2 years old, 95.8 grams, female. As the session concluded Sean gave Maddie the chance to hold the little saw-whet and release it back to the wild. He also snapped this photograph.

In the car ride home Maddie announced, "Papa, I want to be a naturalist."

That winter Maddie and I read the graphic adaptation of Wilson's memoir, *Naturalist*. In a little journal she kept about the book she wrote, "He changed the way we look at the world." When we were separated during the pandemic we watched the PBS special about Wilson titled *Of Ants and Men* together on Zoom. I gasped with delight when this showed up in the chat box—"He's interested in big questions." Yes he is, Maddie.

Maddie Lindberg, about to release the young owl

these things. Save it, you said: in wildness is the preservation of the world . . . Now in closing this letter, I am forced to report bad news. (I put it off till the end.) The natural world in the year 2001 is everywhere disappearing before our eyes—cut to pieces, mowed down, plowed under, gobbled up, replaced by human artifacts.[11]

To recruit new activists to the cause of conservation Wilson penned *The Creation: An Appeal to Save Life on Earth.* His literary device is an impassioned letter to a Southern Baptist pastor. The letter begins:

Dear Pastor:

We have not met, yet I feel I know you well enough to call you friend. First of all, we grew up in the same faith. As a boy I too answered the altar call; I went under the water. Although I no longer belong to that faith, I am confident that if we met and spoke privately of our deepest beliefs, it would be in a spirit of mutual respect and good will, I know we share many precepts of moral behavior. Perhaps it also matters that we are both American and, insofar as it might still affect civility and good manner, we are both Southerners . . .

Pastor, we need your help. The Creation—living Nature—is in deep trouble. Scientists estimate that if habitat conversion and other destructive human activities continue at their present rates, half the species of plants and animals on Earth could be either gone or at least fated for early extinction by the end of the century . . . Prudence alone dictates that we act quickly to prevent the extinction of species and, with it, the pauperization of the Earth's ecosystems—hence of the Creation.[12]

In his book about Gorongosa National Park, Mozambique, Wilson recounts the story of what is possible when nature gets some help. *A Window on Eternity* is about nature recovering from the human-wrought devastation from a sixteen-year civil war. Unfortunately for the park and its inhabitants, it was located near the headquarters of one of the warring armies and thus became a battleground, and a feeding ground. Hungry soldiers killed many large animals for food and sold elephant ivory for guns. After the war, poachers continued to plunder the park. Losses were staggering. From 1972 to 2001 the number of Cape buffalos fell from 13,000 to 15; wildebeest from 6,400 to 1; hippos from 3,500 to just 44; and the number of lion prides and elephant herds plummeted by 80–90 percent. There were no hyenas and rhinos left. In 2004 restoration of the park began, supported by the local people, the government of Mozambique, and philanthropist Gregory C. Carr. The massive effort involved growing and planting rain forest trees, expanding the park boundaries, importing large animals from South Africa, creating a buffer zone around the park, and creating an education and research center. When Wilson visited in 2011 he reported on Gorongosa's rebirth, "While still below their prewar maximum, herds of grazers and browsers once again swarmed across the savanna and grasslands. They were growing fast. The return to ecological balance struck by the old megafauna was on course."[13]

At the conclusion of the book, Wilson calls for similar levels of attention to other large reserves of land and sea and the linking of these reserves to create wildland corridors, like the Yellowstone to Yukon Conservation Initiative (Y2Y), that would run north to south and east to west. Like his call for half-earth, Wilson acknowledges the reach of his proposal. "It may seem radical but only this level of care applied to large sectors of land and sea throughout the world will save biodiversity."[14]

While some are heeding this call to dedicate more space to nature, ecosystems around the world continue to be destroyed and disrupted. The planet's islands of nature (recall the theory of island biogeography) are shrinking in size and growing more isolated. As the theory predicted, species are going extinct or nearing extinction at alarming rates and the populations of species are collapsing, like they did in Gorongosa National Park. Science writer David Quammen memorably wrote: "Islands are where species go to die."[15]

Yet, as we've seen, the Gorongosa story gives us reasons for hope. With some help nature can restore itself. And scientists now know how much space nature needs for its restoration. It's half. Wilson's half-earth call is echoed by other scientific bodies. The International Union for the Conservation of Nature has proposed that 30 percent of the planet be set aside for nature by 2030, a milestone toward the "larger goal of half the planet protected by 2050."[16] Other echoes are Nature Needs Half and Campaign for Nature, a partnership of the Wyss Campaign for Nature, the National Geographic Society, and more than one hundred conservation organizations around the world.

Here is how Wilson speaks to the how and why of the call for half.

Only a major shift in moral reasoning, with greater commitment given to the rest of life, can meet this greatest challenge of the century. Wildlands are our birthplace. Our civilizations were built from them. Our food and most of our dwellings and vehicles were derived from them. Our gods lived in their midst. Nature in the wildlands is the birthright of everyone on Earth. The millions of species we have allowed to survive there, but continue to threaten, are our phylogenetic kin. Their long-term history is our long-term history. Despite all of our pretenses and fantasies, we have always been and will remain a biological species tied to this particular biological world. Millions of years of evolution are indelibly encoded in our genes. History without the wildlands is no history at all.

We should forever bear in mind that the beautiful world our species inherited took the biosphere 3.8 billion years to build, the intricacy of its species we know only in part, and the way they work together to create a sustainable balance we have only recently begun to grasp. Like it or not, prepared or not, we are the mind and stewards of the living world. Our own ultimate future depends upon that understanding. We have come a long way through the barbaric period in which we still live, and now I believe we've learned enough to adopt a transcendent moral precept concerning the rest of life. It is simple and easy to say: Do no further harm to the biosphere.[17]

Like the successful restoration happening in Gorongosa National Park this will take a heroic and concerted effort by citizens, governments, and private organizations all over the world. The audacious, but essential goal of half-earth Wilson presents, if realized, will enable nature to heal itself.

The other rationale Wilson presents for the audacious but essential goal of half-earth addresses human motivation. He contends, and research supports his view, that people are animated and find fulfillment in embracing goals that are bold, hard to achieve, and if accomplished of great benefit. In Wilson's words, "To strive against the odds on behalf of all life would be humanity at its most noble."[18]

At the end of the conference session, the audience rose in a standing ovation. It spoke of gratitude for what they learned about biodiversity's keystone role in wondrously functional ecosystems and for the invitation from this giant of science to widen their professional gaze, to become advocates for the health of nature as well as for patients. They rose for the man who drew on all his faculties in loving service of the natural world. They rose too, I believe, in response to Wilson's walk to the back of the room, for his gracious response to the woman who raised her hand.

REFERENCES

1. Wilson, E. O. (1994). *Naturalist*. Island Press, p. 15.
2. Hölldobler, B., & Wilson, E. O. (2009). *The superorganism: The beauty, elegance, and strangeness of insect societies*. W.W. Norton & Company, p. xviii.
3. MacArthur, R., & Wilson, E. O. (1967). *The theory of island biogeography*. Princeton University Press.
4. Wilson, E. O. (1985). The biological diversity crisis: A challenge to science. *Issues in Science and Technology*, 2(1).
5. National Academy of Sciences. (1988). *Biodiversity*. (E. O. Wilson, Ed.) The National Academies Press.
6. Wilson, E. O. (1992). *The diversity of life*. The Belknap Press, p. 393.
7. Wilson, E. O. (2002). *The future of life*. Alfred A. Knopf, p. 163.
8. Wilson, E. O. (2016). *Half-Earth: Our planet's fight for life*. Liveright Publishing Corporation, p. 14.
9. Harvard Magazine. (1980, January–February). Resolutions for the 80s. *Harvard Magazine*.
10. Wilson 1994, *Naturalist*.
11. Wilson 2002, *The future of life*, p. xxii.
12. Wilson, E. O. (2006). *The creation*. W. W. Norton & Company, pp. 3–5.
13. Wilson, E. O. (2014). *A window on eternity: A biologist's walk through Gorongosa National Park*. Simon & Schuster, p. 8.
14. *Ibid*, p. 141.
15. Quammen, D. (1996). *The song of the Dodo: Island biogeography in an age of extinctions*. Scribner.
16. Dinerstein, E., et al. (2019). A global deal for nature: Guiding principles, milestones, and targets. *Science Advances*, 5(4), p. 4.
17. Wilson 2016, *Half-Earth*, pp. 211–212.
18. *Ibid*, p. 4.

The best arguments in the world won't change a person's mind. The only thing that can do that is a good story.

RICHARD POWERS

A Rewilding Story

TOM BUTLER

Several years ago, my friend Jason was riding his bike. He's an avid cyclist, and on this day he was nearing home after a long ride, going downhill on a paved road outside his small town. Hearing a vehicle coming up from behind, he scooted over as far as he could to the highway's right line. As the van came up alongside, someone leaning out the passenger window screamed "boo" and hit him in the head.

Jason braked hard and swerved onto the shoulder, which was riddled with broken pavement, trying to keep from crashing. Which he did. By luck and skill, he did not wipe out. He did not break bones. He was not killed. But he was righteously furious at the idiots who could have caused him grave injury.

Within a few minutes he arrived home and jumped into his car to scout around town for the van, which he soon found parked outside the local grocery store. Two guys came out. "Remember me?" Jason asked. The first fellow looked blank. "Boo," Jason says. The other guy starts laughing. "He remembers me," Jason said. "I'm the biker you could have gotten killed back there."

A spirited conversation ensued. As the exchange of views escalated, van thug #1 approached with menace, at which point Jason whipped out a wooden axe handle that he'd concealed behind his arm. Swinging the club toward the fellow's head, Jason stopped it just shy of his temple. With a final encouragement to, in all things, but most especially in vehicle/cyclist relations—"BE NICE"—Jason gave the man a light tap on the noggin.

For years thereafter Jason's friends would encourage him to recount the "be nice" tale. No camping trip was complete without it. It became, as friends and family lore often does, part of our collective memory of knowing and loving Jason.

Years went by, and at one point the "be nice" story came up with another friend, John, when Jason wasn't present. Also an avid cyclist, John started telling me about that day when he and Jason had been on a long bike ride together and the van almost ran them off the road. He told how he and Jason had found the van and waited in the parking lot to confront the brutes. And how he watched Jason, with a light tap of his axe handle, encourage those fellows to be nice.

John remembered it vividly; he'd been there after all. And this was curious—because in fact, he had not. Jason was biking solo that day. At the time, Jason and John lived hundreds of miles apart. After years of listening to the tale and admiring Jason's response to the assault, John had simply internalized it. *He had put himself in the story.*

Memory is, of course, highly fallible. Our brains' ability to construct and reconstruct memories is amazing, and not well understood, but recent research suggests that our brains certainly are not flesh-and-blood filing cabinets, Xerox copiers, or computers.[1]

Setting aside the possibility that John had some kind of cognitive impairment (he doesn't), let's switch our gaze from the imperfectness of John's memory to the attractiveness of his delusion.

With each breath, with every heartbeat, we live by grace. But while we live, we organize our lives by stories. We understand our place in the world by the tales we tell ourselves. For as long as our species has employed figurative language, some seventy thousand years, we have been talking and listening, listening and talking, to transmit the wisdom, the humor, the codes of right and wrong conduct that collectively form human culture.[2] Only very recently has this cultural transmission happened through mediated forms of communication. The phones in our pockets, the books on our shelves—that's something new under the sun. For humans, what's tried and true is the oral tradition.

Thus Jason's "be nice" story became a kind of cognitive superglue among his group of friends, sticking together the memories of multiple individuals who were not present at the story's genesis. It was not so surprising, then, that one of us put himself in the tale. Some stories are so attractive that we naturally want to weave our lives into them. They give us meaning. This is most obvious with the world's great faith traditions, which are built on compelling narratives and shape the lives of billions of people. Secular myths also profoundly influence individual and societal behaviors.

If the world humanity is making is based largely on the tales we tell ourselves, do we have the right ones? Or rather, given that cultures around the globe contain many stories that help anchor people to the land and to their wild relatives in the family of life, why is it that the secular myths offered by globalized, techno-industrial culture so dominate our current political and economic affairs?[3]

Even a cursory stab at answering that question is beyond the scope of this essay, but at the very least, we can agree that corporate capitalism offers a very shiny vision of material affluence (and one quite tangible to a small slice of the human population globally), and has a seemingly limitless advertising budget. Alas, that shiny vision is based on a misunderstanding of physical limits and other dangerously false notions that are precipitating climate chaos and unraveling biodiversity. Let's consider just a couple of those problematic notions:

#1: There is something called "nature" and something else called "people" and these are separate, with the former's primary job to serve the latter by providing an endless stream of stuff—"natural resources"—for our use, enjoyment, and profit. This worldview, which is based on a foundational idea of human supremacy so pervasive and unexamined that it's generally invisible, sees the Earth essentially as a large, magical supermarket.[4] That supermarket exists to produce an endless supply of energy, experiences, and material goods for humanity. In contrast, understanding the Earth as a community to which we belong, as one species among multitudes, all sharing a common home and common destiny because all are connected—is more characteristic of grounded, place-based cultures throughout human history.[5,6]

#2: Growth—in human numbers and consumption—can go on, and on. Of this idea, the economist Kenneth Boulding once quipped, "Anyone who believes exponential growth can go on forever in a finite world is either a madman or an economist." It's a good line, but in a sense Boulding was wrong. It's not just crazy individuals or economists who believe, or at least pretend to believe, that this physical impossibility is true. We have based our entire civilization on the secular religion of perpetual growth. Which is, truly, madness.

The result of how many we are and how we occupy the Earth has precipitated the sixth great extinction spasm in the planet's history. A flood of alarming data about crashing wildlife populations and unraveling ecosystems, increasing greenhouse gases, accelerating climate chaos, and growing inequity between the haves and the have-nots in the human tribe are readily available for anyone who cares to look. For those of us who are paying attention to the global eco-social crisis, the fire hose of bad news can be deeply depressing or numbing. But I don't think it's particularly motivating.

So, if the delusional tales we tell are sending us over a cliff, what should we replace them with? What new story is big enough to help turn the trajectory of humanity—and the diversity of life—away from ecological Armageddon? What story is inclusive and attractive enough to inspire millions or even billions of people to put themselves into it?

I vote for this one: the story of *rewilding*, of resurgent wildness enveloping the Earth. Of expanding beauty and diversity. Of wilderness recovery writ large. Of people from all backgrounds and every corner of the globe lending their energies toward helping nature heal, at all scales, to the benefit of all life.

Consider this passage from the *Global Charter for Rewilding the Earth*, drafted for and adopted by the most recent World Wilderness Congress and

Remembering that we are members in the community of life is a first step toward rewilding ourselves, on the path to rewilding the Earth.

endorsed by conservation groups from almost every continent.[7] (There's no "rewilding Antarctica" yet.) The charter's vision statement reads:

> We believe that the world can be more beautiful, more diverse, more equitable, more *wild*. We believe that nature's innate resilience, bolstered by human care, can initiate an era of planetary healing. In that future time when the world is whole and healthy, undammed rivers will run to the sea, their estuaries teeming with life. Following ancient patterns, whales and warblers will migrate unmolested through sea and sky. From tiny phytoplankton to tallest redwoods, all Earth's creatures will be free to pursue lives of quality, and humanity will thrive amidst nature's abundance.

Could this be the dream that's big enough to capture the hearts and minds of millions? That is both timeless and urgent enough to prompt bold action? Could it be the story generous enough to carry our love for specific places into the future, in the form of interconnected ribbons of protected habitat wrapping the planet in wild beauty? Maybe, just maybe it is.

THE COVID-19 PANDEMIC has caused immense suffering, with millions of people dying from a novel virus and billions of others experiencing economic hardship or isolation. If there was the smallest silver lining in a dark pandemic cloud, it was that many people learned, or rediscovered, the joy of being outside... walking, biking, canoeing. In our region, trails were mobbed with hikers. It was tough to find a kayak or snowshoes to buy. People wanted to be outside, in the company of trees and wind and birdsong.

Our bodies and minds are attuned to wild nature.[8] Our stress hormone levels show it. The direct, measurable, physiological effects in the human body of time spent in natural settings as well as psychological benefits are the focus of a fascinating and growing body of research.[9]

While the scientific tools to study these effects are new, the experiential value of time in wild nature is the oldest and most consistent theme in conservation literature. Think of Wordsworth and his cohort rambling about the English lakes district in the late 1700s writing verse extolling the birds and flowers there. Or Thoreau's escaping the already-tamed fields and woodlots of Concord to explore the Maine woods and climb Mount Katahdin in the 1840s. This narrative thread stretches to our day in poetry and prose that helps us relearn kinship with all our relations via reconnection to the beauty produced by "life-seeking creatures in relationship," as Sandra Lubarsky has so beautifully articulated.[10,11]

The desire to reconnect our hearts and minds to the greater community of life, outside and away from obvious artifacts of modernity, drove the first wave of wilderness recreation to the Adirondack mountains of upstate New York after publication of the book *Adventures in the Wilderness* (1869) by Boston clergyman William H. H. Murray. Thus began a period of great popular interest in the Adirondacks, a region relatively little known except by lumbermen who were rapaciously cutting its forests. In nineteenth-century America, forests were fuel. Dead trees meant charcoal to stoke the iron kilns, chemicals to tan leather, saw logs for lumber. The wave of forest-clearing for these purposes as well as agriculture caused the hills and mountains of the Northeast to erode and the rivers to run brown with silt.

So grave was the threat to waterways, which were crucial for transportation and hydro-powered industry downstream, that New York's state legislature created the Adirondack and Catskill Forest Preserves in 1885. Many conservationists had worked for that outcome, but it was no silver

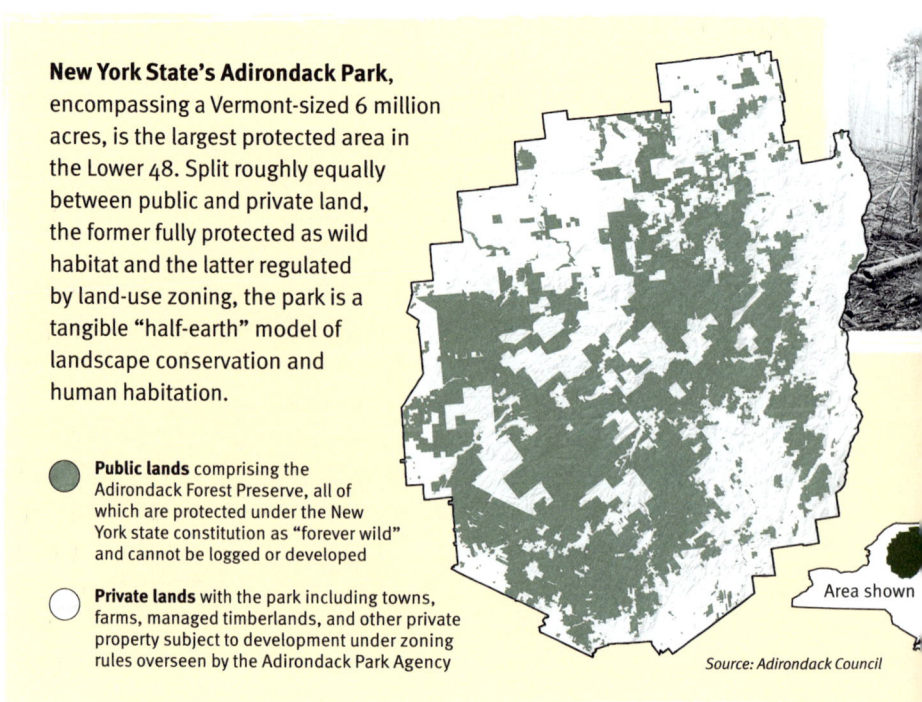

New York State's Adirondack Park, encompassing a Vermont-sized 6 million acres, is the largest protected area in the Lower 48. Split roughly equally between public and private land, the former fully protected as wild habitat and the latter regulated by land-use zoning, the park is a tangible "half-earth" model of landscape conservation and human habitation.

● **Public lands** comprising the Adirondack Forest Preserve, all of which are protected under the New York state constitution as "forever wild" and cannot be logged or developed

○ **Private lands** with the park including towns, farms, managed timberlands, and other private property subject to development under zoning rules overseen by the Adirondack Park Agency

Area shown

Source: Adirondack Council

bullet to stop the logging. Timber merchants would buy private land, cut it over, and then abandon it to the state for unpaid property taxes. That was the genesis of most property which came into public ownership in the new Forest Preserve, lands of which were to be kept as forever wild.

More substantial protections for the region came in 1892 with the establishment of the Adirondack Park, and the passage of an amendment to the state constitution three years later which gives the public lands within the park, the Forest Preserve lands, the highest level of conservation protection for public lands in the United States. They cannot be sold or logged or mined, etc., without a difficult process of constitutional amendment.

The conservationists who were responsible for creating the park and its central legal safeguard, what came be known as the "forever-wild" clause of the state constitution, included the pioneering civil rights attorney Louis Marshall.[12] Marshall was a leading activist against anti-Semitism, an early board member of the NAACP, and a brilliant lawyer who used the courts to challenge structural racism. He was also a dedicated conservationist and father of Robert Marshall, cofounder of the Wilderness Society.

High Peaks Wilderness, Adirondack Park

The Marshall family and so many other wilderness advocates through the decades blazed a path that still leads toward expanding justice, beauty, and health for the land and all the creatures who inhabit, or visit, the park.[13] The Adirondack Park is arguably the greatest example of rewilding on Earth, the fullest expression of the incremental reforestation of the northeastern United States following logging associated with European settlement.[14] Today the Adirondacks are more ecologically intact, more secure wildlife habitat, and a better canvas for natural processes to create, shape, and sustain biodiversity than other parts of the region. The Adirondack Park also provides tremendous social and economic values, and stores vast amounts of carbon naturally, a key to mitigating climate chaos.

The public lands comprising the Adirondack Forest Preserve are *free*, consistent with the etymological roots of the word "wilderness," to follow their own course. They are self-willed lands, home to self-willed creatures.[15]

Last summer, while camped by a lake in the Adirondack wilderness, I watched a loon chick and her momma. Loons are amazing swimmers but not such easy fliers, at least when they take off. They must run along the water surface to gain speed, flapping hard to become airborne. These loons were training. The little one would scamper across the water just as fast as she could go, with mother chasing behind. And then get tired out and stop. Mother loon would rush toward her baby as if to play tag and the little one would start up again. Over and over they repeated this game, getting the little one stronger for a long migration flight in the fall. It was the cutest thing, that loon chick and her momma, at home, free to be loons, in a place that needs never fear the chain saws or jet skis of the future.

"Conservation" as a set of tools including laws and regulations and practices is a broad term, of course. For the ecocentric wing of the conservation movement, which is focused on saving life's diversity for its own sake and not solely for utilitarian ends (although acknowledging the fact that humanity will not thrive on a dead planet), the idea of freedom is central to conservation. This is not a new idea.

Howard Zahniser, author of the Wilderness Act, the 1964 law that created America's national wilderness preservation system on federal public lands, was a part-time resident of the Adirondacks. He had been introduced to the region by conservationist Paul Schaefer, and the Schaefer and Zahniser families ended up having nearby cabins on the edge of the Siamese Ponds Wilderness in the heart of the park. When he was able to escape from Washington, DC, where he served as executive secretary of The Wilderness

Society, Howard Zahniser spent time at that Adirondack cabin where he worked on draft after draft of the Wilderness Act.

As a Pennsylvania native and lover of the Adirondacks' recovering wildness, Zahniser understood that wilderness could grow as well as shrink—the evidence was all around him—and thus he deliberately used the word "untrammeled" in the law's definition of wilderness.[16] Something that is trammeled is bound or caught; untrammeled is free or unimpeded. The Wilderness Act doesn't contain the words "pristine" or "untouched" because the defining characteristic of wilderness is not virginity but *freedom*—freedom to follow its own evolutionary path.

In the decades following bipartisan passage of the Wilderness Act, citizens successfully lobbied Congress to add millions of acres to the national wilderness preservation system. Today there are more than eight hundred wilderness areas in the system, encompassing more than 111 million acres. The growth of the system is heartening, and arguably one of the most positive results of civic engagement over the past half-century. Notwithstanding this progress, however, wild habitats across North America are typically under threat of conversion or degradation if not formally conserved, and often are isolated islands of intact habitat in a sea of development when they are formally protected. Habitat degradation, fragmentation, and isolation is a recipe for extinction.

As the drivers of biodiversity loss were studied and described by the scientific disciplines of landscape ecology and conservation biology in the latter decades of the twentieth century, it became clear to some conservationists that bolder, science-informed advocacy for wilderness recovery was necessary. It would not be enough to simply protect remaining intact natural areas. Rather, it is vital to restore and reconnect habitats, and at a scale adequate to allow ecologically crucial actors—"keystone species"—to repopulate territory from which they had been eliminated by human persecution or land-use changes that had diminished their habitat.[17]

One of the greatest wilderness activists of those decades was Dave Foreman, founder of The Rewilding Institute and various other groups during his storied career in conservation. He is a walking encyclopedia of wilderness history, policy, and activism. In 1991 Dave Foreman and John Davis and a few others founded the journal *Wild Earth*, which first popularized a term, "rewilding," that Dave had coined.[18]

Foreman recognized the need for a new word that meant wilderness recovery on a grand scale, a scale that would allow keystone species such as wolves, cougars, and jaguars, which roam widely, to reestablish healthy populations throughout their native ranges and thus help to reestablish intact food webs.[19] In the 1990s there was a nascent body of research showing how crucial apex predators are to healthy ecosystems, both on land and in the oceans. That understanding has grown through subsequent decades.

The Adirondack Park is arguably the greatest example of rewilding on Earth.

At the same time, the relatively new field of ecological restoration didn't reflect this emerging science; the papers published in its technical journals typically focused on restoring specific degraded areas, not restoring landscape-scale ecological systems. Thus there was a need for a term that captured the latter, and Foreman's genius for language produced "rewilding." The concept landed like a arrow in the hearts of wild

Northeast Wilderness Trust's Eagle Mountain Preserve, eastern Adirondacks

lovers everywhere. A web search on the word now gets millions of results.

New words inevitably evolve when they become broadly adopted, and rewilding is still used to convey its original meaning—large-scale wilderness recovery—but also to encompass many other types of conservation projects. Some conservationists find this evolution in meaning problematic, arguing a need to "put the wild back in rewilding."[20]

Rewilding—as an idea, a meme—is culturally powerful because in a time of loss and diminishment for the diversity of life, it represents the potential and possibility for *more*: More beauty. More abundance. More equity among all the creatures who inhabit the Earth—including humans, of course, even ones who forget their own creatureliness.[21] Moreover, rewilding captivates because almost any of us can put ourselves into that story, either through individual or collective action.

Nongovernmental organizations (NGOs) around the world are doing just that, many of which have joined together in the recently launched Global Rewilding Alliance. Rewilding projects include efforts to expand tiger reserves in India; plant millions of Scots pines to restore Scotland's Caledonian forest; reintroduce beavers and lynx to parts of Europe where they've been eliminated; create prairie preserves for bison in Colorado; and protect some of the most biologically rich parts of the Appalachian Mountains.

Close to my heart is the work that Northeast Wilderness Trust is doing to protect the ancient forests of the future, help set the stage for missing native species to return home, and let diminished natural processes reassert themselves across the landscape. This kind of collective action to foster rewilding through active and passive means, putting nature's needs first while recognizing how wildly beneficial that is to people, is crucial to ending the cascading crises of climate chaos and biodiversity loss.

In practice, a rewilding approach to conservation includes three elements. I talk about them as Places, Processes, and People.

Places: Permanently protected natural areas such as wilderness areas, national parks, no-extraction marine protected areas, privately owned nature sanctuaries, etc., are the foundation of biodiversity conservation.

Processes: Rewilding can happen through active efforts like removing dams and reintroducing missing species, most especially highly interactive species such as beavers, sharks, and cougars. Rewilding can also happen through passive means, that is, allowing for vegetative succession and other natural processes to produce diversity and complexity over time. The historical exemplar for this, again, is the Adirondack Park.

People: Ultimately, to rewild the Earth, we need to rewild ourselves. By which I mean winning hearts and minds to the great cause of conservation, motivated not only by self-interest but also by love for our wild neighbors and kin in the community of life.

Some groups focus on one or two of these streams of rewilding work, some do all three. An example of the latter is Rewilding Argentina, which has implemented the most comprehensive large-scale rewilding program in South America. Beginning in the mid-1990s, that NGO, birthed by American philanthropists Douglas and Kristine Tompkins and led by visionary Argentine conservationist Sofia Heinonen, has helped transform the great Iberá marshlands region of Corrientes Province from a little-known and highly threatened natural area into a world-class destination for wildlife-watching.

During more than two decades of effort, conservationists gained designation of new provincial and national parks, which are contiguous and managed jointly. Community engagement fostered widespread support for the Iberá Park, which is an economic boon to the region. The team of biologists and veterinarians there have reintroduced missing native species including giant anteaters, pampas deer, collared peccaries, and even jaguars to the marshlands.

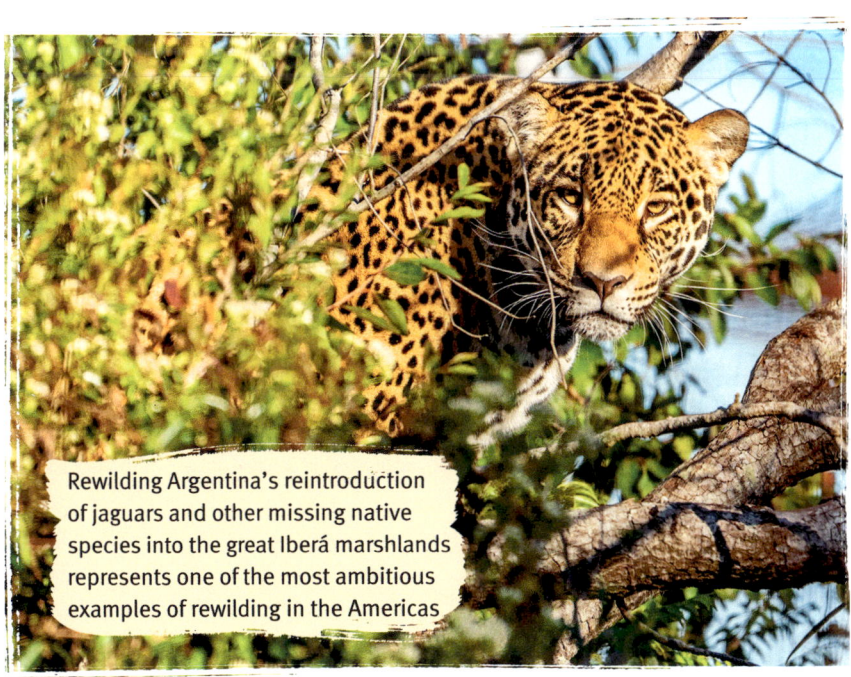

Rewilding Argentina's reintroduction of jaguars and other missing native species into the great Iberá marshlands represents one of the most ambitious examples of rewilding in the Americas

Rewilding aligns perfectly with emerging frameworks for global conservation including "30 x 30," the target for governments around the globe to protect at least 30 percent of their national territory by the year 2030, or the half-earth goal articulated by biologist E. O. Wilson, which mirrors the similarly ambitious Nature Needs Half campaign.[22] These aspirational targets for expanded protected areas address the conjoined climate and extinction crises with the appropriate scale of ambition.[23]

Are they impossible? I don't think so.

With the 6-million-acre Adirondack Park—a park the size of Vermont—we have a model for what a landscape can be, where roughly half is strictly protected as wilderness and the other half is private land managed for farming and forestry, with hamlets, towns, and small cities interspersed in a patchwork of wild and domesticated land. It's not perfect, but it works remarkably well for bears, bobcats, and people. Moreover, courageous conservationists are dreaming about and working to advance progress that will continue and expand the park's role as a global model for integrating human communities and wildlife in a landscape of abundance and opportunity for all.[24]

The idea of blue and green ribbons of wildness knitting up to wrap the globe in beauty is deeply attractive. The many historical and contemporary examples of citizen action that have been successful creating individual building blocks in this future global network should give us inspiration that a future ecological civilization is possible, if we work with urgency and creativity to create it.

Conserving at least half of Earth in interconnected systems of protected natural areas will come to life only through grassroots, bottom-up actions of people, groups, communities, and governments. Acre by acre, parcel by parcel, project by project, rewilding happens when people who love the land work to create the conditions in which nature may rebound.

The question before each of us is: How will we put ourselves into the story of Rewilding the Earth? How will we use our time, influence, energy, and wealth to write that new narrative, centered on beauty, integrity, and reciprocity? How will we help shape a forever-wild future?

REFERENCES

1. Epstein, R. (2016, May). The empty brain. *Aeon*.
2. Harari, N. (2015). *Sapiens: A brief history of humankind*. Harper Perennial.
3. Elder, J., & Wong, H. (1994). *Family of earth and sky: Indigenous tales of nature from around the world*. Beacon Press.
4. Crist, E. (2019). *Abundant earth: Toward an ecological civilization*. University of Chicago Press.
5. Leopold, A. (1949.) *A Sand County almanac and sketches here and there*. Oxford University Press.
6. Kimmerer, R. (2015). *Braiding sweetgrass: Indigenous wisdom, scientific knowledge, and the teachings of plants*. Milkweed Editions.
7. World Wilderness Congress. (2020, April). *Global charter for rewilding the earth*.
8. Wilson, E. O. (1984). *Biophilia*. Harvard University Press.
9. Vedantam, S. (Host). (2018, September). Our Better Nature [Audio podcast episode]. In hidden brain. https://hiddenbrain.org/.
10. Oliver, M. (2005). *New and selected poems: Volume two*. Beacon Press.
11. Lubarsky, S. (2014). Living beauty. In G. Wuerthner, E. Crist, & T. Butler (Eds.), *Keeping the wild: Against the domestication of earth*. Island Press.
12. Graham Jr., F. (1978). *The Adirondack Park: A political history*. Alfred A. Knopf.
13. Terrie, G. (1997). *Contested terrain: A new history of nature and people in the Adirondacks*. Syracuse University Press.
14. Butler, T. (2015). Protected areas and the long arc toward justice. In G. Wuerthner, E. Crist, & T. Butler (Eds.), *Protecting the wild: Parks and wilderness, the foundation for conservation*. Island Press.
15. Vest, J. (1985). Will-of-the-land. *Environmental Review*, 9(4).
16. Zahniser, E. (1992). *Where wilderness preservation began: Adirondack writings of Howard Zahniser*. North Country Books.
17. Noss, R., & Cooperrider, A. (1994). *Saving nature's legacy: Protecting and restoring biodiversity*. Island Press.
18. Johns, D. (2019). History of rewilding: Ideas and practice. In N. Pettorelli, S. Durant, & J. Du Toit (Eds.), *Rewilding*. Cambridge University Press.
19. Foreman, D. (2004). *Rewilding North America: A vision for conservation in the 21st century*. Island Press.
20. Carver, S., & Convery, I. (2021, June 14). Time to put the wild back into rewilding. *Ecos*, 42(3).
21. Berry, W. (2000). *Life is a miracle: An essay against modern superstition*. Counterpoint.
22. Dinerstein, E., et al. (2019, April 19). A global deal for nature: Guiding principles, milestones, and targets. *Science Advances*, 5(4).
23. Locke, H. (2015). Nature needs (at least) half: A necessary new agenda for protected areas. In G. Wuerthner, E. Crist, & T. Butler (Eds.), *Protecting the wild: Parks and wilderness, the foundation for conservation*. Island Press.
24. Adirondack Council, & Goren, J. (2021, November 15). *VISION 2050*. https://www.adirondackcouncil.org/page/vision-2050-332.html (November 24, 2021).

Every natural action is graceful.
RALPH WALDO EMERSON

Conservation in Vermont

Wildness to Devastation, Opportunity to Intention

ELIZABETH THOMPSON

As a teenager, I loved being in the woods. I felt at peace there, far from the noisy crowds at school. The woods were quiet, but there were also signs of a once busy place, of fields and plows and grazing animals. I saw stone walls, abandoned roads, and old cellar holes wherever I went. And the trees were small. This was a new forest, grown up on former farmland. My home in eastern Massachusetts was only a few miles from Henry David Thoreau's cabin on Walden Pond. The woods he knew in the early 1800s were even younger than the ones I knew in the 1970s.

Thoreau wondered what he wasn't seeing. He wrote: "No one has yet described for me the difference between that wild forest which once occupied our oldest townships, and the tame one which I find there today. It is a difference which would be worth attending to."[1]

On returning from the wilds of Maine, where he saw some very old forests, he wrote about that difference: "[The Massachusetts forest] has lost its wild, damp, and shaggy look; the countless fallen and decaying trees are gone, and consequently that thick coat of moss which lived on them is gone too. The earth is comparatively bare and smooth and dry."[2]

Bare and smooth and dry. That's how I remember the forest floor near my childhood home.

When I started working as an ecologist in Vermont, I sought out old forest. I wanted to know: What was "that wild forest?" What might it have looked like before Europeans settled and "tamed" the land? Gifford Woods State Park in Killington, Vermont, was among the places I explored that first summer, and when I saw it, I was absolutely awestruck.

It was like nothing from my teenage rambles. I saw huge trees, but also small and medium-sized trees—trees of many different ages. I saw openings in the woods where trees had died and fallen over, and those spaces were filled with young new growth. The ground was uneven, with high mounds and deep pits resulting from centuries of trees falling over. I saw mossy rotting logs on the ground, with new seedlings growing from them. This forest had a "wild, damp, and shaggy look."

Stimulated by Thoreau's questions about the original New England forest, many scientists are researching what we now call "old-growth forest"— woods that have been shaped only by natural processes for centuries.

David Foster sees old-growth forests as "messy, chaotic, in seeming disarray." He is Senior Conservationist at the Harvard Forest in Massachusetts. Tony D'Amato, of the University of Vermont, likens an old-growth forest to an unkempt garden late in the growing season. Like other conservation leaders, they feel that we need this disarray, that we need wild forests, alongside the younger forests that give us lumber, firewood, and other forest products.

Mature forest, Bridgewater, Vermont

"We are missing many features from our landscapes that only come with very long periods of forest growth," said David. "These include the big live and dead trees, the immense windthrow mounds and downed trees, and the uneven ground that develops over three or four centuries." Old forests, even those that have not achieved the truly old-growth character that takes centuries to develop, offer unique habitats for wildlife. "A big fallen tree can have tip-up roots where winter wrens can nest," explains Vermont Fish & Wildlife's Bob Zaino. "Black bears can den in hollow logs, whether standing or on the ground."

There's also the incredible resilience of these forests. "Old forests are a testament to nature's ability to self-organize, sustain, and rejuvenate in the face of a constant battery of disturbances such as wind, ice, fire, insects, disease, and drought," says forest scientist and conservationist Ed Faison. For instance, the death of a large canopy tree promotes the growth of young trees in the new patch of light, and the fallen trunk becomes a nutrient-rich site for future seedlings.

Old forests provide clean water, clean air, and carbon storage. We're learning that old forests, with their large trees, huge fallen logs on the forest floor, and massive underground root systems, hold more carbon than young ones. Old-growth forests are part of our landscape heritage. David Foster explains that they made up the New England landscape for 10,000–15,000 years, until European arrival. They are now "one of the rarest kinds of forests found here," he says. They're biologically important for their structure and habitat, but these forests are "also emotionally and spiritually important to us," he adds. As author Barbara Kingsolver said, "People need wild places. . . . We need to be able to taste grace and know once again that we desire it."[3]

The clearing of the land, and an explosion of green

> *We have now felled forest enough everywhere, in many districts far too much. Let us restore this one element of material life to its normal proportions, and devise means for maintaining the permanence of its relations to the fields, the meadows and the pastures, to the rain and the dews of heaven, to the springs and rivulets with which it waters down the earth.*[4] —GEORGE PERKINS MARSH

Until about 13,000 years ago, the area we now call Vermont was covered with a mile-thick glacier for thousands of years, leaving the land bare and uninhabitable. Humans returned to the land perhaps nine thousand years ago, and they have affected Vermont's natural communities in many ways. Archeological evidence points to localized heavy use of some of the more fertile valleys, especially near lakes and waterways, by Native Americans since they first arrived here. But their influences on the vegetation were probably minimal in comparison with those of European settlers.

In 1750, Europeans considered Vermont a wilderness, and only a few hardy white settlers had explored the region to trap furbearing animals. The land was probably about 95 percent forested. The only open areas were mountaintops, shores, occasional Native American settlements, and wetlands.

A changing landscape

The New England landscape was, until the seventeenth century, dominated by forests that were shaped by natural processes. As the land was transformed by European settlers, the forests were largely cleared, and by the mid-1800s, half or more of the region had been converted to open agricultural land. As quickly as the land was cleared, it was abandoned, and it is now largely forested again.

The three images above represent a single place in western Massachusetts. The first shows an old-growth forest in 1700. The second, from 1740, shows an early settler clearing the land for a small farm. The third view is from 1930, by which time the farm had been abandoned and was left to regrow to forest.

The chart to the right shows these changes in the land, state by state, from 1600 to the present. *Source: Wildlands and Woodlands: Farmlands and Communities (Harvard Forest, Harvard University)*

European settlers came in earnest in the late eighteenth century, following the end of the French and Indian Wars in 1763. The early nineteenth century saw the clearing of Vermont's forest, with the timber used in lime kilns, converted to charcoal for use in iron and copper production, burned to produce potash, and exported to the south for lumber and a variety of other uses. By the 1850s Vermont was far from the wilderness it had been only one hundred years earlier; nearly three-quarters of the state was cleared, the streams were full of silt from the eroding land, and sheep grazed almost every hillside. In fact, Vermont was for a time the world's largest exporter of wool. When the fertile lands of the Midwest opened up, however, farmers left

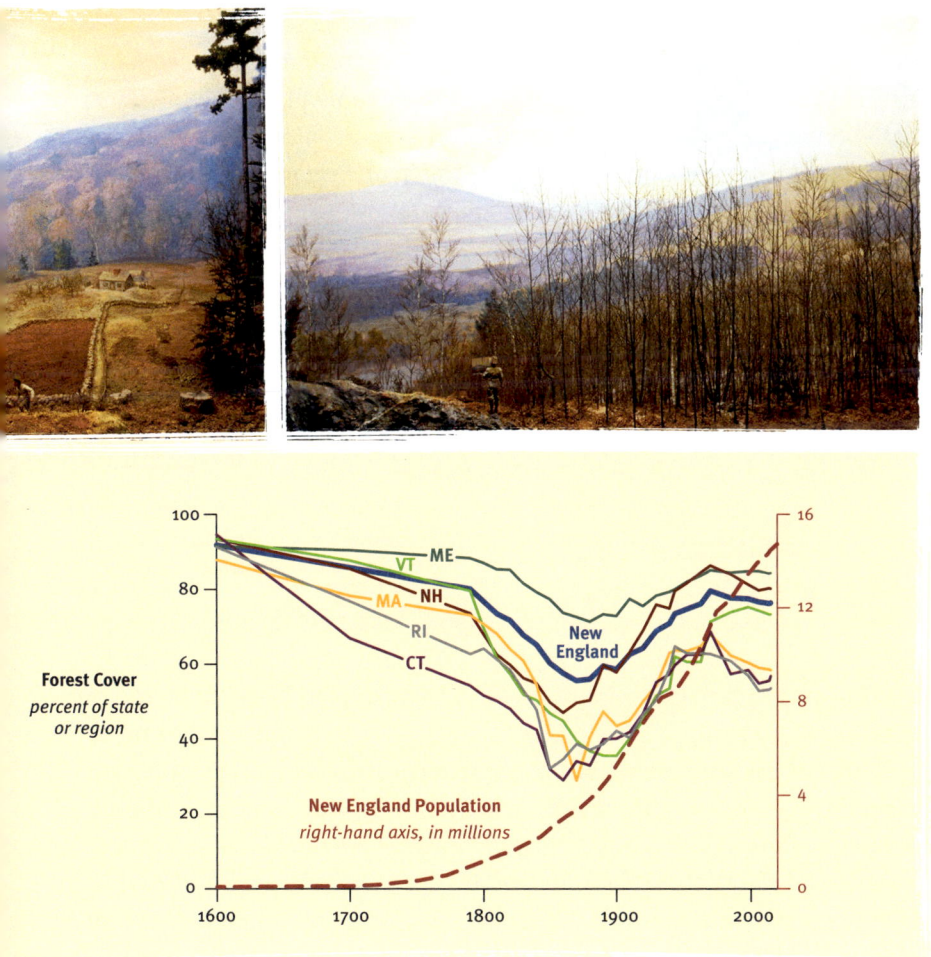

Vermont's infertile hills for the tallgrass prairies and their deep fertile soils. By 1880, dairy farming had replaced sheep, and dairy farming remains an important part of Vermont's agricultural economy. The hill farms of the nineteenth century have been abandoned, and today only about 20 percent of the land remains open. The changes over these short 250 years were dramatic.

I moved to Vermont in 1981, having spent a few years on the wild coast of eastern Maine after college. Coming from rough and rocky eastern Maine, I found Vermont to be tame and bucolic, with its neat villages, grassy hillsides, contentedly grazing cows, and forest-clad mountains. What I soon learned in my hikes and rambles and research was that this bucolic scene concealed the destructive past that Marsh described, and that I now understand—the massive clearing, the severe erosion, the silting of waterways. As I began to study forests and their soils, I began to see things: a young forest with simplified structure and species composition, gullied streams instead of forest seeps, and very little topsoil. In some places, invasive species—honeysuckle, buckthorn, and barberry—dominated the understory. This simplification made it hard to know the "real" Vermont, to understand the natural communities that had been there for millennia before the heavy population explosion brought on by European settlement.

Early conservation efforts: the "leftovers"

We have been here 3 weeks. Have had a fine time. Quite a quantity of fish and berries of all kinds—raspberries, blueberries and blackberries. The logs were all away here the first of August. The steamer only made a few trips while we were here . . . The screeching of a wildcat has been heard several times and Ralph saw it once and shot at it. The boys have gone to the foot of the pond . . . to fish and to get their father who is coming on the 4 o'clock train.[5] —MRS. CHARLES LORD, 1908

One of the places I've always loved since moving to Vermont is Groton State Forest, in Vermont's "Northeast Kingdom." This place reminds me of the wilds of eastern Maine, of the rocky terrain and the abundant bogs. Recently I have made several visits to Peacham Bog, an ecological treasure within that forest. The spongy, mossy ground, the pitcher plants and cottongrasses, and the stunted black spruce reveal a rich geological and ecological history with little human influence. It is a place to feel that grace that Barbara Kingsolver describes.

Groton State Forest is a much-beloved recreational area with some of the state's finest natural features, including Peacham Bog, Owls Head, and Osmore Pond. Its logging history, as encapsulated in the quote above, belies its unique geology. The state land occurs almost entirely on a granite pluton, and the soils are infertile compared to those of the surrounding landscape. The area produced considerable timber in the early days of White settlement, but agriculture never took hold there.

Like the Groton granite, other places that were unsuitable for sheep grazing, haymaking, and crop raising became our state parks, national forests, and other protected public lands. Many of these places—mountaintops, cold boreal cliffs, bogs, and high-elevation lakes and ponds—are stunningly beautiful and have high biological diversity. But other places of high biodiversity fell by the wayside in this "take what's left" approach to conservation. The lowlands, the enriched slopes, the clayplain forests, the calcareous lakeshores, the rich fens—these were not even well known, much less protected, as natural areas early on. They were the places to raise sheep and grow crops.

Looking outside of Vermont we see other, more dramatic examples of this same phenomenon.

Peacham Bog

A rare opportunity

Conservation opportunities are not always about "the leftovers." In 1998, having returned only days earlier from a two-day tour of Champion International Paper Company's 132,000 acres of intensively managed forestland in northeastern Vermont, I received a call from the tour's host. Almost in tears, he reported that Champion would be selling the land.

Vermont ecologists had known for years that this land harbored many treasures—boreal bogs, rare Canada jays and other birds, and old-growth forest remnants. We knew that this was an opportunity not to be missed.

Conservationists sprung into action and, through intense planning and negotiations over many months, conserved the land in a mix of state, federal, and private ownership that provides protection for sensitive features and a 12,500-acre core ecological reserve, along with recreational resources and areas of actively managed timberland.

Yellowstone National Park was an early conservation success in this country, and its protection benefited from the same phenomenon as Groton State Forest—it was spectacularly beautiful, but with its steep cliffs and gorges, and its harsh soils created by thermal phenomena, it was unsuitable for agriculture or, for the most part, even forestry. This lack of "value" was one of the things that convinced Congress to establish Yellowstone as a national park.

In the East, later efforts established national forests that mostly occupied mountaintops and other infertile lands. The Green Mountain National Forest is a legacy of this tradition.

Opportunity to intention

In 1974, The Nature Conservancy (TNC), then a national (now an international) organization, hired its first Science Director, Bob Jenkins. Jenkins had noted that the conservation of places like Yosemite and Yellowstone was based on their natural beauty, and also a ruggedness that made those places unsuitable for agriculture or extractive uses. Said Jenkins:

> I rejected the ideas of saving prettiness, open space, and the like and decided that, in the abstract, there were two worthy objectives for natural land conservation. You could either seek to preserve ecological function, which I called carrying capacity, or you could try to preserve the full array of biological and ecological entities, which I called natural diversity. To have a meaningful impact on the first, I thought, would take an enormous effort, beyond the reach of a tiny conservation organization. Therefore, it seemed to me that [we] should seek to provide ecological lifeboats to save biological species and communities from extinction . . . My original vision was simply to accumulate a knowledge base of constantly updated information about the biota and ecosystems as a basis for conservation action.[6]

During the 1970s, '80s, and early '90s, Jenkins made that vision a reality. He transformed the methods by which TNC established priorities for protecting land and guided the creation of a biological inventory of the Americas, introducing heightened scientific rigor to the group's land-acquisition choices.

I had the pleasure of meeting Bob Jenkins early in my career, when I served The Nature Conservancy's Maine Chapter. I was asked to produce a "biodiversity scorecard" for the state of Maine, and Bob guided me in this effort. The scorecard, created before personal computers and spreadsheets, was a hand-lettered chart of all the significant biodiversity features then known to scientists in Maine: a page for eagle nesting sites, a few pages for rare plant sites, some bogs, a handful of jack pine barrens, and so forth. This was just what Bob had in mind—an inventory of "ecological lifeboats."

By the time I finished graduate school at the University of Vermont, in 1984, technology had already advanced some and the Vermont Natural Heritage Program had just been established, one of the outcomes of Jenkins's vision of having scientists in every state who counted and catalogued the species and natural communities most in need of conservation. I went to work for the program then, as botanist, natural community ecologist, and data manager. Computers were new, and it was all very clunky, looking back, but we persevered, and it worked. We created a database, a technical leap from hand-scribbled charts.

But even before this, the idea of protecting "lifeboats" had already been playing out in Vermont for more than two decades.

Hub Vogelmann, a professor at the University of Vermont and an important early mentor to me and so many others, could easily be considered the father of conservation in Vermont. He started the Vermont Chapter of the Nature Conservancy and conducted the first inventories of natural areas in the state. He published two important books, *Natural Areas in Vermont* and *Some Ecological Sites of Public Importance*, in 1964 and 1969, respectively. These were the first such systematic studies in the state, and they provided some of the very first data for the Vermont Natural Heritage Inventory.

In its infancy, we at the Vermont Natural Heritage Inventory gathered information from the Vogelmann reports and other sources and began doing new inventories based on the knowledge of local botanists, birders, and other naturalists. We combed the museums for clues about where to find rare plants and animals, and then scoured the mountaintops, hillsides, bogs, and swamps to find more. It was an exciting time of discovery.

About a decade after the Inventory was established, a few Vermont conservationists realized that while excellent data was being gathered, there was still not much rhyme or reason to how it was being used in conservation decisions. Conservation of some important natural features was happening, but a lot was still happening opportunistically.

These conservationists, from The Nature Conservancy, the Vermont Agency of Natural Resources, the University of Vermont, and Middlebury College, and a few federal agencies, launched the Vermont Biodiversity

Project in 1995. The project culminated in 2002 with the publication of *Vermont's Natural Heritage*, a guidebook to biodiversity in Vermont, and with the dissemination of a statewide database on priority conservation areas. These areas were identified based on biological "hotspots" as well as enduring physical features that were as yet under-represented in conserved lands.

Notably, this was the first planning project for Vermont or the region that included *enduring physical features*, rather than just biological features. Enduring physical features are the mountains, cliffs, wetlands, and soils that support the plants, animals, and natural communities that make up a place. Scientists who study changes in biota over millennia tell us that we need to be protecting more than just the species we see today, but the landscape on which they evolved, and through which they have moved, over time. These enduring features will last much longer than the present ecological communities, especially as the climate changes ever more rapidly.

Learning to protect enduring features was the beginning, in a sense, of tackling more seriously the work of protecting not just ecological lifeboats, but also *ecological function*, a goal that Jenkins had considered beyond the reach of The Nature Conservancy of the 1970s, a "tiny conservation organization." Protecting enduring features, or "conserving the stage" is now a keystone of conservation planning in The Nature Conservancy and in many state-level efforts such as Vermont Conservation Design. Ultimately, TNC took a national lead role in promoting the conservation of enduring physical features and has continued that work today.

Vermonters worked in parallel with the national TNC efforts in the ensuing years. In 2007, The Vermont Fish & Wildlife Department with the Vermont Land Trust conducted a statewide analysis of significant habitat blocks, which has served as the basis of conservation planning efforts since that time.

Current Vermont intentions

Vermont Conservation Design (see "A Vermonter's Guide for Protecting Biodiversity," page 184) is, at this writing, the most current nature conservation plan for the state of Vermont. It was prepared by a team of scientists who aimed to draw on the best conservation science of the time, and to prepare a plan that would move us through some of the biggest challenges we faced and continue to face.

Among these challenges—all linked and interacting—are climate change, habitat fragmentation, habitat loss, and invasive species.

Vermont Conservation Design draws on the earlier efforts, including the 2002 work of the Vermont Biodiversity Project, the 2007 habitat block project, The Nature Conservancy's Resilient and Connected Network, the international Staying Connected Initiative, and other related efforts. VCD continues to evolve, and most importantly is made available to planners at the local, regional, and statewide efforts through BioFinder, an online planning tool.

The Design uses a three-tiered system of analysis and planning.

At the *landscape* scale, it identifies large, connected features—naturally vegetated habitat blocks, enduring features, and riparian systems—that, if protected, will allow natural communities and native organisms to adapt, move around, and thrive as the climate changes.

At the *natural community* scale, Vermont Conservation Design identifies goals for the protection of all natural communities and the physical settings and ecological processes that support them. Natural communities are repeating assemblages of organisms, their physical environments, and the natural processes that influence them. In addition to their own intrinsic value, natural communities can serve as a coarse filter for the conservation of biodiversity. As an example, if a number of good examples of black spruce woodland shrub bog are protected, there is a good likelihood that the very rare dwarf mistletoe will be protected as well. Vermont Conservation Design identifies restoration goals for some of the features that were greatly reduced with European settlement, such as the old forests described at the beginning of this chapter, and complex young forests resulting from natural disturbance.

At the *species* scale, Vermont Conservation Design identifies those species that will not be adequately protected through the conservation of landscape-scale features or natural communities. Some species have such specific habitat needs, or are so rare, that they may fall through the cracks of these so-called coarse filters. Many of these species are protected through Vermont's Endangered Species Law, and a very few are protected through the Federal Endangered Species Act.

It is our hope that Vermont Conservation Design, as it evolves and as it is put into practice, will inspire other statewide or local efforts to plan for conservation, borrowing concepts and data from the large regional and national efforts by The Nature Conservancy and others, but bringing them closer to home.

Natural communities are repeating assemblages of organisms, their physical environments, and the natural processes that influence them. Examples include Black Spruce Swamp (pictured), Subalpine Krummholz, and Northern Hardwood Forest. All natural communities are worthy of protection, and Vermont Conservation Design sets numerical conservation targets for each of Vermont's 97 natural communities.

The spiny softshell turtle is a species that cannot be adequately protected through the coarse-filter approach. It needs targeted protection of its very specific habitats, the shale beaches where it is known to nest, and even seasonal intervention to protect the eggs from nest predation.

In my own wanderings in Vermont in recent years, I am brought back to Thoreau's musings and I am moved by all the efforts to understand and protect the things that make this place special. Many people are wondering about that "wild forest which once occupied our oldest townships" and are seeking and protecting the remnants of that wild forest. I recently visited an ancient forest in the Green Mountains that, newly discovered, is being protected through local efforts. In the eastern part of the state, an old forest is being protected because the landowner wanted to keep it wild.

Old forests in our future

Vermont Conservation Design (VCD) calls for the protection and restoration of old forests, which once were dominant on the New England Landscape. VCD calls for about 9 percent of Vermont's forest to be designated as old forest. This goal mirrors the earlier regional Wildlands and Woodlands, Farmlands and Communities goal of about 10 percent of the forest as "wildland," or unmanaged forest that will become old growth over time. These old forests will complement other forests in a variety of successional stages, including some very young forests. Achieving old forest goals will require a variety of strategies, from passive management to some restoration and repair.

It is my hope and vision that Vermont will continue to be a place where forests remain forests and bogs remain bogs; where people continue to love and care for those forests and those bogs, and where children grow up knowing that there is a future for them.

REFERENCES

1. Thoreau, H. D. (1864). *The Maine woods*. Princeton University Press.

2. *Ibid*.

3. Kingsolver, B. (2003). *Small wonder: Essays*. Harper Perennial.

4. Marsh, G. P. (1864). *Man and nature: Or, physical geography as modified by human nature*. C. Scribner.

5. *Ibid*, p. 5.

6. Jenkins, R. E. (2010, February 2). Acceptance speech 2010 NatureServe Conservation Award (video recording). YouTube. https://www.youtube.com/watch?v=3ABduiyv5Ic

A wild forest confronts us with what we have done. It reminds us of what we have lost. And it gives us a vision of what—in some way—might live again.

KATHLEEN DEAN MOORE

Adventures in Forest Carbon

Conservation, Carbon Capture, and Our Climate Future

ANNIE FAULKNER

New England is a land of forests. We New Englanders depend on the woods in many ways. Places to work, walk, and recreate. Clean water, warmth, and materials for our homes. Wildlife, biodiversity, natural processes, wilderness. My father was a woodsman, as were his father and grandfather before him. They loved the woods and practiced "good forestry" as it was understood at the time. I inherited their love of the woods and a sense of responsibility for the forest's well-being.

As a citizen and parent, I am deeply concerned about the urgency and magnitude of the climate crisis. As a forestland owner with climate on my mind, I want to know what constitutes "good forest management" in the twenty-first century. As a lover of wild places, I want to understand whether our forest conservation practices are helping the climate or making it worse. Are New England's forests doing as much as they can for carbon mitigation? Should carbon capture and storage be considered a "highest and best use" of forests for the next 30, 50 or 100 years?

These are utilitarian questions, yet I believe forests have their own intrinsic value, too. I have worked for two decades preserving wild places for their own sake. In asking climate questions, my instincts tell me that these twin goals—biodiversity protection and climate stabilization—might lead us in the same direction.

In the search for answers, I was reminded that everything to do with forest carbon is connected to everything else. And in the process, I found myself transformed. Everything I looked at and thought about was colored by carbon.

New England through carbon-colored glasses

During the recent pandemic, I explored the carbon literature, and found my way through forests familiar and new. You could say I bushwhacked toward the answers.

Atmosphere. The first thing I saw with my new carbon lens was actually invisible: the clear blue sky where carbon is rapidly accumulating. The last time Earth's atmosphere held this much carbon dioxide (410 parts per million) was 3 million years ago, when global temperatures were 3°C–4°C (5.4°F–7.2°F) warmer than now, and sea level 80 feet higher.[1] The danger is both invisible and frighteningly real.

My father, Kim Faulkner, loved to split wood.

The source of all this carbon is no mystery. Burning of fossil fuels is the main culprit. Some also comes from deforestation, both past and present. Carbon dioxide is released when forests are cut, burned, or blown down. After a disturbance, forests release more carbon into the atmosphere than they are able to absorb, and this goes on for decades. Disturbed forests become a carbon source rather than a sink, even while the forest is regrowing. It can take a century or more for a forest to recapture all the carbon that was released.[2]

Once upon a time in New England, the land was mostly covered with ancient forests, storing untold tons of carbon. After European colonization, most of that forest was cut down. Scientists estimate that only 40 percent of the carbon that was lost to the atmosphere has been recaptured as the forests have recovered over the past 150 years.[3] As I walk in the woods wearing my new carbon-colored glasses, I think about the other 60 percent of the formerly terrestrial carbon, still up in the air, waiting to get back to the ground.

Trees. On my carbon adventure I learned that trees everywhere—in my backyard, in a friend's woodlot, in your town forest or local conservation area—are about half carbon by weight. Powered by sunlight, their solid, graceful forms grow literally out of the air. It is both simple and miraculous: carbon grows on trees and trees (millions of metric tons of them) grow right out of thin air.[4] In a mixed-age forest, the biggest trees hold the largest share of carbon.[5] It is easy to think of trees as the most amazing carbon-capture and storage technology ever.

Underground. The carbon-colored glasses let me see underground, too. Trees store some carbon below ground and share it with their offspring and other neighbors.[6] Often, half or more of the carbon in a forest is stored below ground, in roots, soil, and vast fungal networks. As forests mature, more carbon is stored underground. The next time you walk in an older forest, you might tune in to the vast carbon pool under your feet, where half of the forest carbon story is being told.

Storing the carbon. My forester friends can estimate the value of a forest in terms of timber stocking, measured in board feet or cords of wood. The carbon stock of a forest is measured in terms of carbon density—metric tons of carbon per hectare (denoted MgC/Ha, megagrams of carbon per hectare).

Across New England, above-ground carbon density varies from the highest amounts in southern parts to the lowest in the far north. Differences in carbon density are attributed primarily to different harvesting regimes,

Where is carbon stored in a forest?

A forest stores carbon in different pools, and the amount of carbon in these pools changes over time.

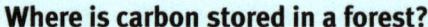

Five forest carbon pools

A Live aboveground
trees, shrubs, and other plants

B Live belowground
roots

C Deadwood
standing dead trees (snags) and downed logs

D Litter
leaves, needles, and small branches

E Soil organic matter
organic material in the soil, such as dead and decayed biomass (e.g., plant material and insects)

Factors that influence the amount and proportion of carbon in each of these pools:

- the age of the forest
- the species of trees making up the forest
- natural and human disturbances
- soil characteristics (e.g., texture and drainage)
- past agricultural land-use history

Source: Forest Carbon: An essential natural solution for climate change (University of Massachusetts Amherst)

with the most intensive harvesting occurring in northern Maine.[7] Carbon density also varies by forest type, with most of the difference in below-ground carbon.

Not surprisingly, carbon density is highest in older forests. A study across northern New England found above-ground carbon averaging 195 MgC/Ha in mature forests and 266 MgC/Ha in old-growth stands. Carbon density was highest where the dominant trees were in the 350–400-year range, and the researchers concluded that maximum density in our region could reach as high as 450 MgC/Ha.[8]

What does this mean for the future of carbon storage in New England's relatively young forests? Assuming average density at 100 MgC/Ha today, and the maximum potential at 250–450 MgC/Ha, New England's forests could, over time, capture and store *two to four times more carbon* than they do now.[9] In place of the usually gloomy climate news, I found the magnitude of this untapped carbon capacity incredibly hopeful.

Carbon capture. If old forests and old trees store the most carbon, which trees or forests capture the most carbon annually? Conventional thinking is that young stands grow quickly, forest productivity peaks and declines, and old stands lose as much carbon as they gain. From this perspective, forest ecosystems become "overmature" and eventually become carbon-neutral.[10]

Current carbon science tells a different story, however, for both individual trees and forest stands. Carbon capture is determined by leaf area and, as the tree grows, carbon uptake actually accelerates.[11] While a young tree or forest will grow at a relatively fast rate, the volume of carbon it captures annually is fairly small. In contrast, an older tree or forest might grow at a slower rate but, in the absence of a major disturbance, it will pack on increasing amounts of carbon year after year as total leaf area expands.[12]

The ongoing and impressive growth of old trees is confirmed in a global study of over six hundred thousand trees from over four hundred tree species. The study concluded that rapid growth in giant trees is, in fact, the global norm, not an exception limited to a few species.[13] To someone who marvels at really big trees, this was wonderful news.

Researchers at Harvard Forest came to a similar conclusion for older forest stands. In a forest characterized by 80–120-year-old hardwoods, the rate of carbon capture doubled between 1992 and 2015.[14] A separate, global study of forests more than 180 years old found that net carbon storage was positive in three-quarters of them.[15]

These studies and others challenge the conventional wisdom about the carbon capture and storage capability of bigger trees and older forests. A recent synthesis put it this way:

> It is now clear that trees accumulate carbon over their entire lifespan and that old, wild forests accumulate far more carbon than they lose through decomposition and respiration . . . This is especially true when taking into account the undisturbed soils only found in unmanaged forests . . . The concept of overmature forest stands . . . does not apply to carbon.[16]

Bringing carbon home

What does all this science mean for landowners and others who care about forests? Can forest management decisions here in New England contribute meaningfully to global carbon mitigation? Helping forests "be all they can be" for the climate is fundamental to so-called "Natural Climate Solutions," which involve protecting and restoring natural carbon sinks, including forests, farmland, grasslands, wetlands, peatlands, mangroves, and more.[17,18] Already, forests recover and store approximately 14 percent of U.S. emissions.[19] On top of that, Natural Climate Solutions may have the potential to sequester an additional 21 percent of the country's net emissions.[20]

Among the natural carbon sinks, scientists estimate that forests have by far the greatest potential to increase sequestration.[21] At an average of about 80 percent forest cover, New England is the most forested region of the country and is located within one of the world's largest intact temperate forests.[22] And, at an average age of 75 years, New England's forests are poised for a period of rapid carbon accumulation if allowed to grow.[23] This gives me confidence that forest management decisions here could have global climate implications.

How, then, should New England forestland owners (my family included) work with forests to tap into their potential and bring more carbon home? Forests are connected to so much of what we care about, including and beyond the climate, that there is no one-size-fits-all answer. Conversations around climate, carbon, and forest management generate numerous options, well-informed opinions, and understandably strong feelings. Here are just a few ideas that I've explored on my forest-carbon adventure.

No net loss of forests. At the most basic level, it seems obvious: we should protect all of New England's remaining forests as forests. This means no more "terminal harvests" that convert forestland permanently to non-forest uses. As we do with wetlands, maybe "no net loss of forests" would be acceptable, such that if some very small amount of deforestation is unavoidable, an equivalent or larger area is restored to forest. Additionally, we should avoid any harvest of New England's remaining old-growth forest. Though small in acreage, these special places are unparalleled for carbon storage, ecological history, and rich biodiversity.

After 150 years of forest recovery, New England is again losing forest; approximately twenty-four thousand acres are lost annually to development.[24] In addition to protecting existing forests, we can also encourage reforestation, especially in the least forested parts of southern New England.

Do it in the backyard (or back forty). For those fortunate to own homes and perhaps an acre or more, we can take inspiration from Connecticut resident Ed Faison, who took a careful look at the trees, shrubs, and lawn right in his yard.[25] A scientist himself, Ed calculated the carbon costs and benefits of various landscaping options. To enhance carbon capture and storage at home, he lets almost all of his trees grow old, especially the largest ones that store the most carbon, and he leaves dead trees standing when safe to do so. Ed moves downed trees and branches to brush piles or scattered in the woods rather than chipping them. He removes invasive vines that choke some trees and has stopped mowing part of his lawn to

Forest-related actions—including reforestation, avoiding deforestation, and low-impact forest management—have significant global potential for natural climate mitigation.

allow it to reforest naturally. He calls these practices an "exercise in restraint" and embraces the natural beauty and enhanced diversity that result from leaving more vegetation intact. Ed's "backyard climate solutions" can be applied to school grounds, a business campus, a town park, or any smaller parcels.

Wildlands as strategic carbon reserves. Landowners eager to prioritize long-term carbon sequestration and storage can do so with a conservation easement. A "forever-wild" designation (also called "ecological reserve," "natural area," or "wilderness") can be included in an easement if requested by the landowner.

I was lucky to be raised in a family that valued wilderness along with "good forestry." When my father's generation conserved the family forest with an easement 30 years ago, they designated roughly a quarter of it as wilderness (we call it Wildcat Hollow). It was an unusual thing to do.

Nowadays, wilderness-style conservation easements are more familiar, if still atypical. Vermont-based Northeast Wilderness Trust holds "forever-

Wild forests are the hardest-working forests. —TOM BUTLER

wild" conservation easements on over twenty-seven thousand acres in New England and New York. For land under a more conventional "working forest" easement, managing a portion as wildland is nearly always compatible with existing restrictions and with existing working forestland tax deductions (and is as easy as not mowing some of the lawn).

Exploring new conservation areas this past year, I wondered how many landowners think about carbon at all. Land trusts and other nonprofit landowners often manage some or all of their property as natural areas and some are making these designations permanent. The Vermont chapter of The Nature Conservancy recently dedicated their five-thousand-acre Burnt Mountain Preserve to carbon storage in perpetuity. A forever-wild easement with Northeast Wilderness Trust ensures the property's future as an old-growth forest. Middlebury College has placed a similar conservation easement on its twenty-one-hundred-acre Bread Loaf property with Vermont Land Trust, maximizing carbon storage over time.

Town land can also function as a carbon reserve. The town of Whately, Massachusetts, is working with the Kestrel Land Trust in nearby Amherst to dedicate a new town forest to wilderness and carbon. Any town forest protected as wildland, such as in my old hometown of Stoddard and nearby Chesterfield, New Hampshire, can function as part of a regional forest carbon reserve.

It is interesting to consider what role state lands should play in carbon mitigation. Vermont's Agency of Natural Resources manages over 350,000 acres of land, for example, including some areas designated as "forever-wild." Should state lands be enrolled to help meet carbon mitigation and biodiversity conservation goals? At present, only 3 percent of New England is protected strictly for nature and natural processes.[26]

Letting today's natural forests continue to grow old without interference is sometimes referred to as "proforestation."[27] I wonder what the carbon impact would be of not harvesting for several decades (or more). The Nature Conservancy's new-and-improved Resilient Land Mapping Tool helps answer this question. For a woodlot, a town forest, or a prospective conservation project, one can estimate a baseline carbon stock as of 2010, and potential additional carbon sequestration to 2050, assuming no harvesting or other major disturbance.

Naturally, I wondered what proforestation could do for carbon on the land my husband and I own. With information provided by the Monadnock Conservancy (who holds our conservation easement), the Resilient Land tool estimated 52,000 metric tons of carbon on the property as of 2010 (around 95 tons/acre). For 2050, the estimate is almost 59,000 metric tons (around 105 tons/acre), for an increase of about 6,000 metric tons (about 10 percent). These volumes are on the low side, but not surprising. When we adopted it, the forest was in poor condition—degraded and just beginning to recover.

I also wondered, what would a half century of no harvesting mean for forest carbon in all of New England? It turns out that researchers at Harvard Forest have modeled this very thing: forest carbon storage in New England under various scenarios beginning in 2010. They estimated that New England forests would add 53 percent more carbon by 2060 if left unharvested (and even more when they included a warming climate).[28]

Because everything is connected, our forest carbon decisions will have other impacts as well. One can imagine how an expansion of wildlands in New England could increase logging pressure elsewhere, potentially zeroing out any net carbon gains. Plus, we New Englanders already consume more wood than we harvest, meaning our impacts already "leak" outside the region. Thinking holistically, a commitment to wilderness may require other commitments, such as reducing wood consumption and increased wood recycling. Reduce, Reuse, Rewild.

Carbon-conscious forest management. Beyond wildlands, carbon mitigation requires us to think about forestry practices and carbon dynamics on the region's remaining forests. Currently, one sees a range of management regimes, from short-rotation intensive harvesting for pulpwood and biomass (more common in northern New England), to long-rotation forestry aimed at large sawlogs and high productivity, and more. Across the region, few forests are managed optimally for carbon sequestration and exploitative practices are common. A recent study found almost 40 percent of northern New England forests significantly degraded in terms of stocking, species diversity, and overall productivity.[29] At the same time, many family forest owners across New England, with their foresters, manage in a way that does accumulate carbon, even if that is not their objective.

Fortunately, guidance is available for landowners wishing to improve carbon mitigation while keeping the door open for future harvesting. Vermont Family Forests offers a checklist of optimal conservation practices that explicitly protect carbon sequestration and storage in actively managed forests.[30] Maine author Mitch Lansky promotes a "double bottom line" approach to both increase carbon sequestration from forest management and reduce carbon emissions from forest products.[31] There is also "Exemplary Forestry," a set of guidelines from New England Forestry Foundation (NEFF) to simultaneously increase carbon sequestration and storage, timber quality, and forest productivity.[32]

The basic strategy of carbon-conscious forestry is straightforward: decrease harvest frequency, and leave more large trees behind, alive and dead. This gives middle-age and older trees more time to pack on the carbon, protects belowground carbon pools from disturbance, and leaves dead trees to slowly decompose. Also: design harvests to diversify the age and size structure and species composition of the forest.[33] Forest structures that favor carbon sequestration, biodiversity, and other ecosystem values also favor volume production.[34] There is further advice to protect soil carbon, including restricting activity to dry or frozen ground conditions and specific

Examples of carbon-conscious forestry practices

- Grow the largest trees and use the longest rotations possible within site and log quality limitations.
- Forests generally sequester and store the most carbon when left untouched. Avoid creating canopy gaps other than those that are deemed essential to meet non-ecological forest functions and values.
- Use single-tree and small-group selection methods. Avoid clear-cutting and whole-tree harvesting.
- Biological legacies of the forest community (coarse dead wood, logs, and snags; trees that are large, living, and old; buried seeds; soil organic matter; invertebrates; sprouting plants; and mycorrhizal fungi) should be protected to aid in post-harvest recovery and to keep the forest from becoming "oversimplified."
- Log under frozen, winter conditions to maximize the soil's ability to store carbon. Delay summer harvests. Avoid spring harvests and deep rutting.
- Take special care to protect wetlands, particularly those with muck and peat soils and a thick organic layer. These wetland soils are often capable of storing ten times as much carbon as other soils in the region.

Source: Vermont Family Forests

slopes, minimizing rutting, and avoiding wet or peat areas, among others.

Does forest management to increase forest carbon actually work? The answer appears to be yes, at least in some circumstances. A study in Vermont found that, ten years after treatment to foster old-growth conditions, experimental stands held much more carbon than conventionally managed stands.[35] Similarly, using Exemplary Forestry, stocking on NEFF's properties increased dramatically, despite ongoing harvesting. According to NEFF, their forests are older, more diverse, better at sequestering and storing carbon, and more resilient to climate change, and they produce more and better-quality wood compared to typical industrial forests.[36]

One of the hurdles for the landowner pursuing carbon-conscious forestry is the "transition phase"—the years when some income is deferred while the forest matures.[37] For landowners and public land managers who don't need immediate income from their forest, longer-term, carbon-conscious management plans can be a boon for forest carbon and result in a more valuable forest product down the line. A reduced harvest plan can ensure that carbon gains are not lost in future logging.

What all of this tells me is that forestry practices that increase carbon sequestration and storage can provide an important public benefit in our climate emergency. So, how could we help more landowners to manage with carbon in mind? One idea is to include carbon standards among the "best management practices" of forest certification programs, such as Current Use programs, the American Tree Farm program, and the Forest Stewardship Council.

Show me the money: paying landowners for carbon. Carbon markets might be another way to advance carbon-friendly forestry. Carbon credits can be generated when landowners prove that they manage their forest in a way that stores and sequesters carbon above a minimum threshold. For landowners, the income generated from the sale of carbon credits can augment timber income and/or defer the need to harvest.

Historically, carbon markets were accessible only to large landowners with long time horizons. Recently, land trusts have developed "aggregation projects" in which multiple ownerships are combined for the purposes of a single deal. The carbon credit industry is also developing less expensive ways to measure carbon on forest land, aiming to open carbon markets to much smaller parcels.[38] And conservation groups are developing approaches in which landowners are paid to implement specific practices with known carbon outcomes.[39,40]

For all their potential benefits, carbon credits are complicated and controversial. Critics question whether carbon projects actually increase net carbon sequestration and/or keep carbon out of the atmosphere for the long term.[41,42] Carbon credits are also criticized for "greenwashing" businesses or giving them a "license to pollute." Some think the entire industry needs a massive overhaul.[43] Closer to home, Northeast Wilderness Trust is trying to address these issues with their "Wild Carbon" credits generated from forests newly protected as forever wild.

In the end, wild forests, carbon-conscious forestry, and carbon markets cannot match the scale and urgency of the climate crisis. Even with the highest-quality forest carbon credits, we cannot sequester our way out of this emergency.[44] Other mechanisms are essential to dramatically reduce emissions as soon as possible.

What next? The path to 2050

The climate emergency demands an urgent, dramatic response. Yet forests live and grow at their own pace. Is there anything we can do right now to accelerate carbon mitigation in New England's forests?

In northern New England, and especially Maine, short-term economics and current ownership patterns deny forests nearly all of their carbon potential. Letting these youngest and most cut-over forests grow into middle and older age could be New England's best opportunity to quickly accelerate carbon capture. For carbon mitigation, the "highest and best use" of these forests may actually be to free them from logging, for 30, 50 or 100 years, and in some places forever.

Trees and forests living today could be the living, breathing, carbon-capture super towers of the future.

In other parts of New England, middle-aged forests with relatively low harvest rates yield positive carbon returns at present. If management regimes were refocused to achieve older forest conditions, and some forests were allowed to rewild, these forests could accelerate carbon mitigation in the 30–50-year time frame, when they reach an even fuller stride in annual sequestration. Assuming we can decarbonize the economy, there will be decades, even centuries of work to decarbonize the atmosphere. Trees and forests living today could be the living, breathing, carbon-capture super towers of the future.

A comprehensive carbon-conscious forest strategy for New England will require an "all of the above" approach. Clearly, we need vast wild lands for biodiversity, carbon mitigation, and for humanity. We also need carefully managed forestland to grow wood for essential needs. We can get there if we let today's older forests grow even older, manage younger forests on longer rotations, and ensure that some forests are never cut again. This will only work, however, if we also lower our demand for and consumption of wood products. A comprehensive forest strategy must address consump-

Young Nina and Ben King with friend Geoff Jones and a big old yellow birch, Stoddard, New Hampshire

tion, so that our local needs are better met locally, and we stop exporting our forest carbon footprint to other parts of the world.

For my family, I imagine what our forest might look like in 2050 or so. Considering the property's condition when we bought it, what would a 30-year carbon-focused management plan look like? The first step might be a forest carbon inventory, including tree size and age structure, species composition, and estimates of above- and below ground carbon stocks. We would want to know how our forest fits in with conservation and management on neighboring properties and in the region. We could even plan our own experiment: leave half to regrow freely and manage the other half to produce old-growth character over time, and hopefully some nice sawlogs by midcentury. No matter what the world looks like then, I can only imagine that a mixture of wild woods and old, productive forest would be the right thing to leave behind—for people, wildlife, and the climate.

REFERENCES

1. Brannen, P. (2021, February). The Earth's terrifying warning lurking in the Earth's ancient rock record. Our climate models could be missing something big. *The Atlantic*.

2. Law, B. E., Thornton, P. E., Irvine, J., Anthoni, P. M. & Van Tuyl, S. (2001). Carbon storage and fluxes in ponderosa pine forests at different developmental stages. *Global Change Biology*, 7(7).

3. McKinley, D. C., Ryan, M. G., Birdsey, R. A., Giardina, C. P., Harmon, M. E., Heath, L. S., et al. (2011, September). A synthesis of current knowledge on forests and carbon storage in the United States. *Ecological Applications*, 21(6).

4. Anderson, M. (2021, April). *Wild Carbon with Mark Anderson*. [Video]. YouTube. https://www.youtube.com/watch?v=cihFT9l0ids

5. Stephenson, N. L., Das, A. J., Condit, R., et al. (2014). Rate of tree carbon accumulation increases continuously with tree size. *Nature* 507: 90-93.

6. Zhou, D., Liu, S., Oeding, J., & Zhao, S. (2013). Forest cutting and impacts on carbon in the eastern United States. *Scientific Reports*, 3:3547.

7. Harvard Forest. (2017). *Wildlands and woodlands: Farmlands and communities, broadening the vision for New England*. Harvard University Press.

8. Keeton, W. S., Whitman, A. A., McGee, G. C., & Goodale, C. L. (2011, December). Late-Successional biomass development in Northern Harwood-Conifer Forests of the Northeastern United States. *Forest Science*, 57(6).

9. *Ibid*.

10. Odum, E. P. (1969, April). The strategy of ecosystem development. *Science*, 164.

11. Stephenson, Rate of tree carbon accumulation increases continuously with tree size.

12. Anderson, *Wild Carbon*.

13. Stephenson, Rate of tree carbon accumulation increases continuously with tree size.
14. Finzi, A. C., Giasson, M. A., Barker Plotkin, A. A., Aber, J. D., Boose, E. R., Davidson, E. A., et al. (2020). Carbon budget of the Harvard Forest Long-Term Ecological Research site: pattern, process, and response to global change. *Ecological Monographs* 90(4): e01423.
15. Luyssaert, S., Schulze, E., Börner, A., Knohl, A., Hessenmöller, D., Law, B.E. et al. (2008, September 11). Old-growth forests as global carbon sinks. *Nature,* 455.
16. Anderson, *Wild Carbon.*
17. Fargione, J. E., Basset, S., Boucher, T., Bridgham, S. D., Conant, R. T., Cook-Patton, S. C., et al. (2018). Natural climate solutions for the United States. *Science Advances* 14: eaat1869.
18. Catanzaro, P., & D'Amato, A. (2019). *Forest carbon: An essential natural solution for climate change.* University of Massachusetts, Amherst.
19. Friedel, M. (2017, July). Forests as carbon Sinks. *Loose Leaf.* American Forests.
20. Fargione et al., Natural climate solutions.
21. *Ibid.*
22. Harvard Forest, *Wildlands and woodlands.*
23. Moomaw, W. R., Masino, S. A., & Faison, E. K. (2019, June). Intact forests in the United States: Proforestation mitigates climate changes and serves the greatest good. *Frontiers in Forests and Global Change,* 2(27).
24. Ducey, M. J., Johnson, K. M., Belair, E. P., & Mockrin, M. H. (2016). *Forests in flux: The effects of demographic change on forest cover in New England and New York.* Carsey Research, National Issue Brief #99. University of New Hampshire, Carsey School of Public Policy.
25. Faison, E. K. (2021). Backyard climate solutions. *Arnoldia* 78(3).
26. Northeast Wilderness Trust, https://newildernesstrust.org.
27. Moomaw, Intact forests.
28. Duveneck, M. J., & Thompson, J. R. (2019). Social and biophysical determinants of future forest conditions in New England: Effects of a modern land-use regime. *Global Environmental Change,* 55.
29. Gunn, J. S., Ducey, M. J., & Belair, E. (2019). Evaluating degradation in a North American temperate forest. *Forest Ecology and Management,* 432.
30. Vermont Family Forests. (2020, March 6). Organic forest ecosystem conservation checklist. [Checklist]. https://familyforests.org/wp-content/uploads/2020/03/Organic-Forest-Ecosystem-Conservation-Checklist-2020-03_06_2020.pdf
31. Lansky, M. (2016). The double bottom line: Managing Maine's forests to increase carbon sequestration and decrease carbon emissions. http://planetmaine.net/meepi/lif/managingforestcarbon.docx
32. New England Forestry Foundation. (2021). *Exemplary forestry.* https://newenglandforestry.org/learn/initiatives/exemplary-forestry/; Final Report—https://newenglandforestry.org/wp-content/uploads/2020/08/Exemplary-Forestry-Acadian-Forest-040919-final.pdf ; Standards—https://newenglandforestry.org/wp-content/uploads/2018/11/NEFF-EF-standards-metrics.pdf
33. Hoover, C., & Stout, S. (2007). The carbon consequences of thinning techniques: Stand structure makes a difference. *Journal of Forestry,* July/August: 266–270.

34. Ford, S. E., & Keeton, W. S. (2017). Enhanced carbon storage through management for old-growth characteristics in northern hardwood-conifer forests. *Ecosphere* 8(4).

35. *Ibid*.

36. New England Forestry Foundation, *Exemplary forestry*.

37. Thom, D., Golivets, M., Edling, L., Meigs, G. W., Gourevitch, J. D., Sonter, L. J., et al. (2019). The climate sensitivity of carbon, timber, and species richness covaries with forest age in boreal–temperate North America. *Global Change Biology*, 25, 2446–2458.

38. Finite Carbon. https://corecarbon.com, Forest Carbon Works https://forestcarbonworks.org/

39. Family Forest Carbon Program. https://www.familyforestcarbon.org

40. Popkin, G. (2020, June 4). How small family forests can help meet the climate challenge. *Yale Environment 360*. Yale School of the Environment.

41. Lang, C. (2020, December 14). *The Nature Conservancy's fake forest offsets*. REDD.

42. Pearce, F. (2021, March 9). Is the "legacy" carbon market a plus or just hype? *Yale Environment 360*. Yale School of the Environment.

43. UCL. (2020, December 4). Carbon offset market needs radical reform. *UCL News*.

44. Waring, B. (2021, April 23). There aren't enough trees in the world to offset society's carbon emissions—and there never will be. *The Conversation*.

You cannot help but learn more as you take the world into your hands. Take it up reverently, for it is an old piece of clay, with millions of thumbprints on it.

JOHN UPDIKE

High Flyers

PRUCIA BUSCELL AND JOE ROMAN

You see a flash of bright yellow in a field, and a small bird with a black cap and white markings on its black wings rises swiftly and takes off in bouncy, undulating flight. It is sometimes called "the potato chip bird" because its high-tinkling four-syllable call seems to say "Potato chip, potato chip." That's how Sandra Fary introduces the American goldfinch, one of the first birds in her ornithology unit at Camels Hump Middle School.

"We learn the easiest bird calls first, the chickadee, the phoebe, the ones you hear clearly and are easiest to identify, then we learn the warblers, which are trickier," explains Fary, a science teacher in Richmond, Vermont. Her students spend six weeks learning about the lives, behavior, and songs of birds. They see pictures, videos, and live birds during school bird walks. They study Cornell University's *All About Birds* website. Sometimes Fary intones fun phrases reminiscent of a certain bird's call. "When kids repeat the phrase 'if I see you I'll seize you and I'll squeeze you till you squirt,'" she says, "they'll recognize the song of the warbling vireo."

One of Fary's favorite science outings is a three-day ornithology field trip to Vermont's Northeast Kingdom, a region with more than two hundred lakes and ponds and undisturbed boreal forests near the Canadian border. "I love that most of the kids know very little about birds when we start," Fary says. By the time they're on this trip, most know 40 bird songs. They can practice recognizing birds by sight, flight, and habitat as well as sound, and they savor those proficiencies long after. "They will often email me that they've heard the song of a white-breasted nuthatch or a tufted titmouse," Fary says.

Fary says her philosophy for her seventh- and eighth-grade students is to let kids experience both hard science and nature beyond the classroom. She organizes as many as 30 scientific outings a year to fields, streams, ponds, and woodlands where they are introduced to ecosystems in water, earth, and air, as well as the chemistry of soils and the practice of data collection. "Sometimes we meet other students, sometimes we meet experts," she says. "We put the experience in the kids' hands—that's important for them."

Fary has taught kindergarten and high school, and when she first taught middle school 25 years ago, she knew it was the place for her. "It's an important time for kids," she says, explaining she loves their untarnished curiosity and receptiveness. "It's an age when they are deciding what they love and what they hate—a lot of big decisions. I feel honored to have a part in that."

Fary's own immersion in nature began in childhood. Her grandfather, a wildlife biologist for the state of California, included Fary and her brother in his summer field research in the mountains. Her grandmother, a self-taught

Before her class paddles out to observe loon families, Sandra Fary reads *Secrets of the Loon* by Laura Purdie Salas.

botanist, had the children identify every tree, bush, wildflower, creature, and cloud formation. "Those things gripped me," Fary recalls. "I knew what excited me as a kid." Now she designs her classes to share that excitement.

Fary believes nature studies need to include a sense of place. In a recent article in *The Journal of Natural History Education*, Matthew Kolan and Walter Poleman, lecturers at the Rubenstein School of Environment and Natural Resources at the University of Vermont, endorse that view, describing the importance of an in-depth understanding of the particular places we inhabit—their unique living organisms, climate, culture, and physical features. They believe learning local conditions on a small, human scale is vital for thoughtful human responses to local conditions and local people. "If we are sensitive to the nuances of place, we can inhabit without destroying," they write.[1]

> *Great fiction is often praised for evoking a strong "sense of place." Birds do the same. In my own backyard, watching the types and rhythms of birds each day and each season heightens my appreciation for the subtler workings of the landscape . . . Seeing the world through the eyes of birds gives me a sense of place like no other.* —CHRIS CANFIELD

Learning about birds, and their nesting, feeding, and flight, offers a glimpse of a region's distinctive environment and biological diversity. Scientists say birds are indicator species that serve as sensitive barometers of environmental health, and their well-documented decline is a danger signal.[2] During the ornithology trip, the students see high-elevation birds that live mainly in spruce forest, then they paddle downriver to learn about shorebirds. "We become birds," Fary says. "We build nests, and the kids have to find material that each bird would use, and find food based on the shape of its beak." The experience gets personal. As Fary stood in a line with students holding dog food in their outstretched hands, a Canada jay hopped from hand to hand eating the treat.

Trish O'Kane, another environmental studies lecturer at the University of Vermont (UVM), also values the relevance of place, and her own awe of locale began with birds. She considers herself an "aviation extension agent" for three months a year—a task and title she created for herself, as she explains in a *New York Times* essay: "I help eastern bluebirds, black-capped chickadees and tree swallows raise their young in five tiny wooden birdhouses in Burlington," she writes. "The rest of the year I teach environmental studies . . . and serve as a freshmen adviser. The jobs are surprisingly similar. In both spheres—the avian and the academic—I work with creatures who make me laugh, make me cry and inspire me not to give up on this troubled world."[3]

In her first years of teaching environmental studies as a doctoral student at the University of Wisconsin, O'Kane observed, "Most of my undergraduate students have spent the bulk of their environmental education inside heated or air-conditioned classrooms. They experience nature through PowerPoint. They can cite the causes of global warming, but they cannot recognize the goldfinch in the bush outside the classroom window or even the robins grazing on campus lawns." She was determined to get them outside.

She designed the class "Birding to Change the World," in which university students pair with elementary school children to learn about birds and other living creatures in their own neighborhoods, in Wisconsin. She resumed the course after coming to Vermont in 2016. For the fourth- and fifth-grade birding class at J. J. Flynn Elementary School in Burlington, O'Kane paired each child with a college mentor by asking both what kind of animal they'd like to be. Big cats were paired with big cats, reptiles with reptiles, raptors with raptors. The pairs would meet once a week and explore what the youngsters most wanted to learn. For some, the flicker of a snake in ferns or the sight of bright red cardinal can initiate a long-term interest.

Kolan and Poleman say relationships are vital to understand and apply nature's lessons. "Social relationships offer the opportunity to deepen our practice of natural history by sharing stories, discoveries, insights, and questions, and holding each other accountable for commitments we make and the edges that we explore," they write. "Some of the strongest learning communities we have encountered are those that create the conditions for personal relationships to flourish."[4] Exemplifying this concept, Fary and O'Kane build teamwork and mentoring into their classes.

Many of O'Kane's young "bird buddies" have retained their avian enthusiasms as well as connections with their mentors. Fary's teaching nourishes camaraderie among students and adults. She also nurtures relationships with naturalists and scholars at Cornell, Middlebury College, and UVM, as well as with representatives of government agencies and public and private organizations. They're the experts and scientists who enrich her students' experiences by joining the outings and sharing their insights, which range from mosses to plate tectonics. Sometimes they help release salmon smolt from hatcheries, or they remove invasive plants, replacing them with native trees.

Study hard what interests you the most in the most undisciplined, irreverent and original manner possible. —RICHARD FEYNMAN

Dressed in cowboy boots and a pair of moose-poop earrings, Fary teaches her students to identify flora and fauna, rocks and trees, and understand the dynamics of rivers, wetlands, watersheds, water quality, and migrating populations.

They learn to be open to the natural world: "Are you familiar with a stinkhorn mushroom?" Fary asks. (Hint: the Latin name is *Phallus*.) "Middle schoolers can't do anything but laugh hysterically if they see a stinkhorn. We came across this old sugar maple that had hundreds of stinkhorns up and down the whole tree. We just sat there, mesmerized and laughing for it must have been half an hour."

Fary's field trips are the stuff of local legend, and kids love to mull them over in anticipation and memory. Every teacher knows field trips are difficult and dicey. While teaching a marshland class to City Corps teens (a precursor to AmeriCorps) at Jamaica Bay in New York City, one of us saw

In Trish O'Kane's "Birding to Change the World" classes, college students join school children to learn about birds and other living creatures in their neighborhoods.

a man running by with a gun, yelling, "Stop or I'll shoot." Heart racing, Joe yelled for his students to hit the ground. And then the New York moment: a film crew emerged from the reeds. "Cut!" the director called out. The students seemed unfazed.

Fary takes her 12- to 14-year-olds on more than 50 trips over two years—in addition to teaching chemistry, physics, and the principles of Concept, Evidence, Reasoning that they will need for state testing.

"I still get nervous before the first trip of the year, and I have an OCD-level obsession going over every detail," Fary says. The trips require extensive preparation, collecting permissions and money, coordination among educators, making arrangements with people at the planned site, and lining up experts and the dozen adult volunteers who supervise and guide 40–50 kids. Are the trips dangerous? Fary grins. "They're not always safe," she says. "But the only time a canoe ever capsized it was the adult canoe." That was during a wetland ecology unit, at a pond in Ferdinand, Vermont, where a large bay merges with fen, marsh, and swamp, a place known for water snakes and strong wind. Some stress, but no calamities. "We've had skunks and raccoons getting into the sleeping tents," she says. "One time a student woke up and screamed when she saw a skunk sitting on the bottom of her sleeping bag gobbling up her bag of Doritos."

Fary also teaches English, using literature to illuminate science, and she fosters careful, accurate writing. Her students read E. O. Wilson's *Letters to a Young Scientist*, the E. O. Wilson Biodiversity Foundation's Half-Earth Project website, and Annie Dillard. She uses enticements to capture each child's imagination—music, songs, rhymes, photography, games, art, and nature design by the English sculptor, photographer, and environmentalist Andy Goldsworthy. Robert Frost's poem "Spring Pools" helps prepare for a trip to a vernal pool where students learned about reptiles and amphibians.

Bailey Willett is an entomologist and research assistant at the Harvard T.H. Chan School of Public Health, where one mission is the global eradication of malaria. It was in Fary's class a decade ago that Willett decided to be a scientist. "She has great intuition about what engages 12- and 13-year-olds," Willett says of Fary, "and there was a great dynamic of respect and responsibility, which was unusual. We were young scientists engaging with the lessons and with experts."

When Fary invited Willett to talk to her class in January 2020, Willett recalled Fary's visual and tactile ways of presenting concepts. So she used layers of Jell-O to create a replica of a human arm and let students "bite" the arm with syringes. Willett wanted students to know that the way mosquitoes get our blood is "more straw than jaw."

Willett has vivid memories of field trips, recalling a nighttime amphibian crossing where frogs and salamanders traversed country roads to get to water. "All of us were wearing reflective clothing, and headlamps, and it was dark, raining and cold, and we were guiding the amphibians across the road, and some we picked up," she recalls. "If we saw cars, we signaled them to slow down so the amphibians wouldn't get hit. It was exciting." After that outing, Willett says, she and her sister often went back to the spot as citizen scientists looking for frogs and salamanders and any environmental changes.

Fary enjoys teaching about amphibians. "Kids come in not knowing much about them, or they have misconceptions," she says.

On one trip to a vernal pool, the middle schoolers were convinced they heard ducks quacking as they walked the trail. As they approached, the pond went silent. "What happened to all the ducks?" they asked. Fary knew how long it would take for the animals to start calling again. "I am the wood frog whisperer," she told them. They rolled their eyes. Then they gathered around and hushed as she read from the book *Life of the Wood Frog*. Just as she got to the part about wood frog voices, as if on cue, the pond erupted in sound. The kids murmured, "Oh my gosh, you really are the wood frog whisperer."

"I wish it were always that perfect," Fary muses.

Tell me and I forget, teach me and I may remember, involve me and I learn. —BENJAMIN FRANKLIN

For the last decade, the North Branch Nature Center (NBNC) in Montpelier has worked with 20 elementary schools to enhance nature learning. Sean Beckett, director of natural history programming at the center, says a current effort is designed to include middle schools and high schools.

Vermont Alliance for Half-Earth, co-founded by Curt Lindberg and several half-earth advocates, including Fary, Katie Sullivan, a teacher at Warren Elementary School, and Charlie Wanzer, a high school math and science teacher at Twinfield School in Plainfield, explored local educational needs and helped create NBNC's Vermont Biodiversity Institute for Educators. Beckett says Fary was instrumental in putting together the institute's recent offering on environmental learning and the art of the BioBlitz.

"Through the half-earth meetings, we realized we wanted biodiversity education for teachers," Fary says. "I reached out to the science teachers in my district. There really was no biodiversity curricula. Eight teachers said they'd take the course. I pitched it to Sean, and he gave us everything we asked." Biodiversity can be incorporated into all aspects of basic biology, evolution, natural selection, or the impact of climate, Fary notes. The institute's first class was held in 2019. It is now a regular summer offering.

The BioBlitz is a biological census. —**NATIONAL GEOGRAPHIC**

In a BioBlitz, students, teachers, community members, and scientists take an inventory of every living organism they find at a chosen site. It can be an afternoon on a school campus, or occur over many days across different habitats. According to E. O. Wilson, the first BioBlitz was held on July 4, 1998, at Walden Pond, in Massachusetts, in honor of Henry David Thoreau.[5] Beckett says BioBlitzes are easier now because of iNaturalist.org, a joint initiative of the California Academy of Sciences and the National Geographic Society. Participants photograph the living things they encounter and upload pictures to the site. Experts help with the identifications, and the pictures contribute to a database of biodiversity in the area.

In 2018, NBNC, along with Montpelier's Community Services Department, and its Conservation Commission, hosted a city-wide BioBlitz. Beckett said 500 people armed with bug nets and binoculars identified 1,600 different species.

Fary helped create Camels Hump Middle School's first BioBlitz, along with 10 other schools in the Winooski Watershed, in 2019. At Warren Elementary School, in Warren, Vermont, Katie Sullivan, who teaches third and fourth grade, organized a BioBlitz for more than a hundred children and teachers in grades one through six. "The kids had nets, bug boxes, vials, other collecting equipment," she says. "They all had iPads, and they were very excited. The kids felt like they really were biologists."

Nathaniel Sharp, a data technician from the Vermont Center for Ecostudies, who provided support for the multi-school BioBlitz, talked to the youngsters afterwards and found them eager to learn more about all the things they had seen. He says all school participants in the BioBlitz identified 735 species, more than one hundred species of mushrooms, and an orange and yellow robber fly rarely seen in Vermont. Sharp himself has been a birder since childhood, when he was inspired by *My Side of the Mountain*, a book by naturalist Jean Craighead George about a boy and a peregrine falcon. That falcon is still his favorite bird, and he thinks birds are often the first interest of people drawn to nature because birds are everywhere, they adapt to nearly every possible place, and their flight and song capture our imaginations. "Nathaniel was a birding mentor in one of my first Birding for Change classes at UVM," Trish O'Kane notes. "Then he helped me develop the class as my teaching assistant—an incredible person."

Sullivan reports, "The children were excited to learn they'd helped paint a picture of what's going on in our little part of Vermont." Like Fary, Sullivan emphasizes hands-on learning—anything the kids can build, test, feel, or eat. In a lesson on scientific method, for example, her children work with mold on bread and on orange peels. "They ask questions about what mold is, they isolate variables, and watch mold grow, and see how their variables made it more moldy or less moldy," she says. "And it's gloriously gross, which 8- and 9-year-olds love." She has nonvenomous corn snakes in her classroom and used to have doves. She'd ask the students whether birds were warm-blooded or cold-blooded, and they didn't know. "If you hold a dove and a snake," she explains, "you feel the difference, and you know."

A spotted salamander gets a helping hand while migrating across a busy road.

> *"Hope" is the thing with feathers—*
> *that perches in the soul—* —EMILY DICKINSON

Sullivan has been teaching for 38 years. Her hope for all her students is a lifelong sense of wonder about the natural world. "I want them to appreciate the beauty of science, and feel empowered to use science to make the world a better place," she says. "I want them to know that they have tools to protect the environment. And if they have early experiences that make them passionate about the natural world, there's a good chance they'll continue to feel that way when they grow up and the concepts and challenges are harder."

In conversation with several educators, hope—for the future, for individuals, for the planet—emerged as a driving force that makes them persist through disappointments and difficulties. Speaking of her students, Fary says, "The hope I have for them is that they develop respect for nature and make smart decisions for themselves and their communities in the future."

"Understanding the natural world, and the microbiological world, offers incredible potential for us and harbors secrets about where we've been and where we can go," says Wanzer, the high school teacher. Experiences that instill expertise and love of science, he adds, can help keep young people from falling prey to the common notion that science is only for elites wearing lab coats. Willett agrees. "Access to the outdoors makes a difference in how people perceive science," she says, adding that educators can emphasize citizen science and science jobs that don't need advanced education.

For O'Kane, studying nature means the hope to instill the practice of thinking deeply about human rights in relation to the way we treat the planet and each other.

> *All birds, of course, are miracles, and humans have known this for millennia. We have looked to birds as oracles. Our hearts soar on their wings and their songs. Even the tiniest bird can teach us that life is larger than humankind alone.* —SY MONTGOMERY

In 2005, Trish O'Kane was living in New Orleans and teaching journalism at Loyola University. In August, Hurricane Katrina devastated the city and surrounding areas, causing 1,800 deaths and billions of dollars in damage. She and her husband were among the evacuees. Their one-story home was under 11 feet of water for three weeks. "And it wasn't just water. It was a toxic soup, electronic stuff, oil spills, sewage, garbage," she recalls. "Everything was destroyed. They were searching the neighborhood for bodies."

The experience was life-changing. O'Kane thinks of her life as pre-Katrina and post-Katrina: "I realized the environment and social justice are connected—you can't separate the environment from racism and poverty . . . I wanted to learn everything I could about nature."

"I hadn't previously been interested in birds," she says. "But I wondered. Where did birds go in a hurricane? How did some survive? I didn't have a plan. But I was outdoors, observing birds. I was feeding those little house sparrows." Bird populations initially plunged in New Orleans, and birdsong was disconcertingly absent. Birds did soon return to this broken, poisoned place, and O'Kane was intrigued and uplifted, even by the non-native sparrows that had sheltered in place during the storm.

In O'Kane's earlier career, she was an advocate for social justice and a journalist investigating human rights issues. In Guatemala, where she lived for six years, she investigated massacres and human rights abuses. The author of *Guatemala: The People, Politics, and Culture*, O'Kane immersed herself in the culture, living for a time in isolated mountainous areas. "People there had resilience and great respect for the Earth," she says. After returning to the United States, she documented hate crimes and abuses against marginalized people for the Southern Poverty Law Center, taught journalism, gave writing classes for women in prison, and wrote articles for several U.S. publications. The environment hadn't been her focus. Hurricane Katrina's devastation pushed her in a new direction.

"We are such indoor animals," she reflects. "I realized I had been disconnected from place. We need to be connected with what is happening outside." She thought about the sense of place among Mayans in the Guatemalan highlands and in the lives of her parents, who came from Ireland. "My father loved being outdoors. He grew up on a farm, where he loved the birds and wee creatures, and he knew all the wells, springs, water and rocks. He was very conscious of place, and I grew up with that."

The importance of physical locale infused her academic work at the University of Wisconsin–Madison, where she earned her doctorate in environmental studies.

She was living on the edge of Madison's 213-acre Warner Park. On one side, in an affluent neighborhood with spacious homes, green space was ample. On her side, where more people of color and less affluent people lived, the land was targeted for development that would bring more concrete, lights, and noise, with plans for a huge parking lot. Examining a general question of green space access, she saw in that microcosm of inequality a common pattern. She and her husband started an environmental defense project, learning the history of local wetlands once used as dumps and organizing nature clubs for kids, efforts she describes as "environmental education on steroids."

Attending hours of public meetings, O'Kane realized that council members genuinely thought their parking lot plan would help the poorer community. "They didn't see all the life in that park. All the animals in the park and children playing there were invisible to them," she recalls. "When I told people I was birding in the park, they were shocked. They didn't think it was safe."

O'Kane began a citizen-science bird survey in the park with her undergraduate college students, middle school students, and neighborhood residents, with support from the local Audubon Society. They studied migratory birds, including gray catbirds, that fly thousands of miles from Wisconsin and the northern United States to Central America or the Caribbean and back—a lesson in the interconnection of local and distant places.

When a middle school principal asked for help on a youth program, O'Kane suggested a birding club with kids and college students. One goal was cultivating the kind of citizenship that makes people revere the natural world and want to protect it. "I love how birds can be our teachers," O'Kane observes. "Birds have fascinating ways of solving problems. Can we learn from them? And listening to birds on an early spring morning is a lesson in magnificence—hearing the fluty ethereal notes of the wood thrush and the wren's joyful, irreverent song."

> *Birds matter, not least, because amazing migrations remind us what an interconnected web we live in, from pole to pole.* —BILL McKIBBEN

Adults often overlook environmental issues that seem remote and abstract, but teachers say youngsters grasp the importance of biodiversity and climate change. They get the concept of a warming planet, and the danger of extinctions, and the entwinement of local and global ecosystems. But those understandings aren't automatic. Kolan and Poleman urge educators to shift away from didactic instruction to what they term "design for emergence."

Yellow-billed cuckoo

Emergence, they explain, is the process by which new patterns, ideas, and structures arise in complex systems like schools and communities. In psychology, that means the development of human consciousness from interactions among billions of brain cells, and in ecology emergence explains the coordinated behavior of schools of fish and flocks of birds. In education, emergence means creating the conditions for learning to take place, rather than prescribing outcomes.

Field trips and birding classes are examples, as are Fary's short "sit spots." Students choose a spot outside, under a tree, in a field, or on a bench in a garden, she explains. "They stay still for five minutes, in a place in contact with nature, and observe everything—sights, sounds, smells, textures, their emotions. They can write it all down or do sketches."

Concepts of environmental planning stretched when Fary had students use maps for the *Global Climate Change and Human Impact* unit. They used half-earth maps of New England, Vermont, their towns, or their own yards. They collaborated to find out how the land was used, and to identify population densities, and housing, agriculture, and business. They had to figure out how much land was preserved already and choose new areas

Students gather on the banks of the Winooski River to explore floodplain dynamics.

of potential conservation to reach the ideal of 50 percent preserved. A final version of the maps will be published for use in middle schools nationally. Fary's eighth graders also built a bird-friendly pollinator garden on school grounds to attract birds, bees, bats, and butterflies.

Wanzer, the high school teacher, thinks all science students would benefit from robust instruction in math, chemistry, and physics, and more reading and note-taking. But he also immerses students in nature and history. Eastern pearlshell mussels in the White River watershed, for instance, actually do make valuable pearls. Treasure seekers plundered their populations in the 1800s, and more recent deforestation and agricultural runoff have hit them hard. They can live from 80 to 90 years, Wanzer notes, and their unusual reproduction depends on unrelated creatures. Males release their sperm into the water. Females filter it, fertilize it internally, then produce larvae, which they eject into the water by the thousands. To survive, larvae must attach themselves to the fins and gills of trout and other cold-water fish, where they remain briefly before dropping into the sand. Like birds, the pearlshell mussels are an indicator species for climate change, Wanzer says, because they need cold water to reproduce. If the water gets too warm, there are no trout.

Fary says her principal strongly supports the field trips she considers so essential. Her advice to teachers dealing with less supportive administrations and tight budgets? "You can find the money," she says. "There are grants. We have fund-raisers. Take advantage of the opportunities out there. There are conservation organizations, colleges, nonprofits. Get to know them. Network with other teachers. And educate yourself on what you teach and what you love." Fary takes classes and attends lectures whenever she can, from the local Audubon center to studying ornithology at Cornell.

The 2020 pandemic was frustrating for educators across the country— time lost in childhood is never fully regained. In-person classes were suspended for months, and teachers faced the challenge of online instruction without a playbook. It was heartbreaking for Fary. She lost her teaching team, the day-to-day interactions, and all the field trips she had planned. How to do place-based teaching during a lockdown? She encouraged her students to get out in their yards and identify birds and plants as part of the spring BioBlitz. Some of them thrived. Others did not. Fary was anguished to realize that some of her students, through stresses beyond their control or hers, lost interest in school. Nature experiences can relieve some of the pressures that lead to childhood depression.[6] Yet the balm of nature, and the stories that emerged from the field trips, were mostly lost to this group.

Faced with this sadness, Fary designed her own nature challenge—to visit every lake in Vermont where she could find a common loon. When pressed, she'll confess that it's her favorite bird. "I'd thought it would be simple," she said, but it turned into a beautiful and unusual experience. "I spent time talking to the locals on these ponds and lakes. I was invited to stay for dinner—and this was during the pandemic—eating outside, of course, talking about loons. I loved what I learned and I fell in love with loons." She will never forget the haunting call of a loon she heard over a lake on a summer evening.

O'Kane also found respite in birds, just as she had in New Orleans. "I had needed something to make us happier after Katrina," she recalls. She watched wrens, starlings, hummingbirds, chickadees, and catbirds. She studied birds that migrate between the American Midwest and Northeast and Latin America, much like her own journeys. She loves birds and their stories. She'll tell you about Wisdom, the world-renowned 70-year-old Laysan albatross who is still hatching chicks, and the pet European starling that

Luna moth

captivated Mozart with such a gift for vocal mimicry that the composer thought he was hearing his own music.

"And chickadees," O'Kane continues, smiling as she recalls the chickadee fledglings that clung to her hand when she held out some seed. "Two of them just stared at me ignoring the seed. They're curious. They're tiny, and spunky . . . and they are really ferocious little things." They're smart, endearing, and vulnerable and sometimes pesky, like the freshmen she'll advise.

On the last day of Fary's eighth-grade class, a luna moth appeared in the school's pollinator garden. Luna moths, with their pale green wings, are a rare sight, especially during the day. Their adult lifespan is only between five and seven days long, and they're usually nocturnal. "It was as though the moth was posing for us, as though we had come full circle," says Fary, who happened to be wearing her luna moth T-shirt. "And when we left, it flew away. That must have been its last day of life. It was found dead on the playground the next day." It is now in a display case in Fary's classroom for future students to see.

REFERENCES

1. Kolan, J., & Poleman, W. (2009). Revitalizing natural history education by design. *The Journal of Natural History Education*, 3(32).
2. Fitzpatrick, J. W., & Marra, P. P. (2019, September 19). Crisis for birds is crisis for all. *The New York Times*.
3. O'Kane, T. (2019, September 5). Of fledglings and freshmen. *The New York Times*.
4. Kolan, J., & Poleman, W. (2009). Revitalizing natural history education by design. *The Journal of Natural History Education*, 3(37).
5. Wilson, E. O. (2006). *The Creation*. W. W. Norton & Company.
6. Bratman, G. N., Anderson, C. B., Berman, M. G., et al. (2019). Nature and mental health: An ecosystem service perspective. *Science Advances*, 5 (7).

*To climb these coming crests
one word to you, to
you and your children:*

*stay together
learn the flowers
go light*

GARY SNYDER

Saving the Forest by Learning the Trees

ALICIA DANIEL

In wild Vermont places where children see blue herons and chase leopard frogs, where mosses snug down tightly over rocks in quiet streambeds, where people can lay hands on maple trees and feel the sweet spirit flow like sap, falling in love with nature just happens. It happened to all of us writing this book. Perhaps that is the easy part. Harder is to take that love and turn it into action on behalf of the wild. All of us are trying to do that, too. I do it through my work as a Field Naturalist for the City of Burlington, working to protect and connect wild lands. I do it by teaching the next generation of naturalists both in the University of Vermont Field Naturalist Program and in the Vermont Master Naturalist Program. By teaching, I am returning a gift that was given to me.

Because no one becomes a naturalist by accident. High school guidance counselors don't suggest the career "naturalist" as a good fit based on an aptitude test. The first spark is a fierce desire and then it becomes a path you have to blaze. I started on that path in the summer of 1987, when I accepted one of four spots in the Field Naturalist Master of Science Program at the University of Vermont (UVM). I left my research position with the Texas Senate Committee on Natural Resources, packed up my Honda Civic, and drove the 2,300 miles from Austin to Vermont. So vast is Texas's sense of its own place in geography and history that I had to look Vermont up on a map. Closing the *physical* distance between where I was and where I wanted to be turned out to be the easiest part.

Four years since its inception, UVM's Field Naturalist (FN) Program was still experimental; a colt finding its legs. Its founder, Hub Vogelmann, was looking for bold people. People with "moxie." Looking back, that drive across the width of the country was a bold move—some might say desperate. I was an *amateur* naturalist, volunteering on bat cave surveys and birding at the local sewage treatment ponds—our inland Texas sea for shorebirds including long-billed dowitchers and glossy ibises. Oh, and I was an English major entering a Master of Science program. But the pull to be a naturalist was irresistible. When a friend sent me a *New York Times* clipping on the FN Program, I started salivating. Suddenly I knew that all I had ever wanted to be was a field naturalist. Even though I had never met one—or even *heard* of one.

A charismatic, larger-than-life figure, Hub created the FN Program in his image. He designed the program with a seemingly preposterous ambition: to educate the next generation of Charles Darwins, John Muirs, and Rachel Carsons. People who could take the conservation issues of the times by the horns. People who could—and would—change the world. Through his research on acid rain, he'd already helped pass the federal Clean Air Act of 1963—and seven years later, he helped to author Vermont's Act 250, a monumental piece of legislation that mandated broad protections, especially for high elevation forests in the state. He'd also founded the Vermont Chapter of the Nature Conservancy. Hub chewed over the idea of an FN Program while he hiked up Camel's Hump to sample the acid rain combed out of the clouds by red spruce and balsam fir needles. He refined the design while he paddled a canoe around Shelburne Pond, wooing wealthy donors to conserve its watery beauty. Finally, he launched the FN Program in 1983 from his office at UVM where he was Chairman of the Botany Department (still quaintly "chair*man*" not "chair" of *Botany* not Plant Biology), an office that looked out across the UVM green to Lake Champlain and the Adirondacks beyond.

After I graduated from the FN Program in 1989, my own path led me to found the Vermont Master Naturalist (VMN) Program in 2016. Coincidentally, I was the same age as Hub when he founded the FN Program. In many ways, the VMN Program is designed to bring a slice of the FN Program to the people. So what do the Field Naturalist Program, a rigorous two-year Master of Science Program, and the Vermont Master Naturalist Program, a nine-month professional certification program, have in common? What are these programs contributing to the conservation of biodiversity in

Vermont and beyond? And, most importantly, how can you help?

Both programs teach naturalists to look at the big picture. Many naturalists are specialists, who view the world through the focused lens of a birder, wildlife tracker, geologist, or botanist. And, yet, setting out to be a naturalist by learning the *pieces* can be so daunting. (Herpetologist Jim Andrews jokes that he picked his field because there are only 40 species of salamanders, frogs, toads, snakes, lizards, and turtles in Vermont.) In many fields, naturalists can never know all of the *pieces*—and if you doubt me, think mushrooms, think mosses, and then think soil microbes.

Field Naturalists and Vermont Master Naturalists do set out to learn the pieces, but more importantly they learn to see the patterns and understand the processes that shape a landscape. They can recognize red pine, and also know that a stand of red pine planted in rows in sandy soils in Vermont is likely a soil conservation project dating back to the 1930s and 1940s. A stand of red pine on an otherwise oaky knoll often tells a story of lightning strike and fire. Pieces, patterns, and processes is a framework. Once naturalists learn scientific frameworks, they can use them to make sense of any place in the world. Besides pieces, patterns, and processes, one of my favorite frameworks for training naturalists in a holistic way starts with dessert. It starts with the layer cake.

The Vermont Master Naturalists of the Mad River Valley among old trees on the Catamount Trail in Lincoln Gap before graduating, June 2021.

The layer cake

Imagine slicing through a landscape like the layers of a cake. Then consider all the interactions between the layers. How does geology influence soils, which in turn set the stage for plants? How do plants create habitat for animals? Where does water fit into the picture? Laid down on its side as if on a plate, the layer cake also becomes a timeline of Vermont's natural and cultural history. The rocks in Vermont are hundreds of million years old. The surficial geology above them—the tills, gravels, sands, and clays that are parent material for our soils—derives from glaciers that scraped and scoured their way across Vermont and are less than thirteen thousand years old. Above this layer, we find the early and generally light influences on the land from native people fishing, hunting, and farming over twelve thousand years. And written over that, the evidence of the cycles of the clearing and regrowth of Vermont forests by European settlers over the last 250 years leading into the modern forests and human habitations of today.

Based on work of the Scottish landscape architect Ian McHarg, University of Pennsylvania, the layer cake approach arrived in Vermont in the 1980s, after UVM Field Naturalists started to study it in the summers under Tom Siccama (Yale) and Art Johnson (UPenn) at the Yale School

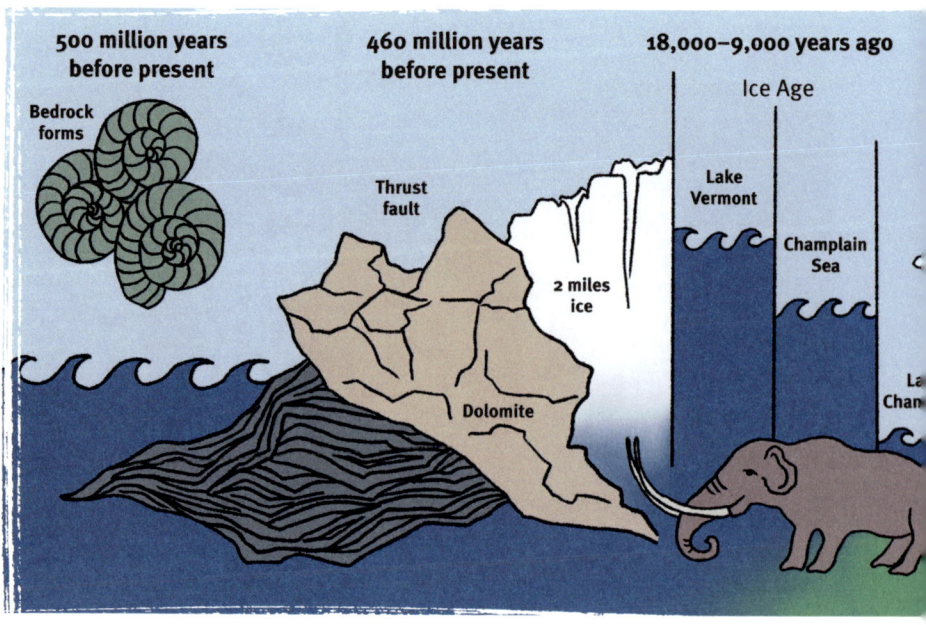

of Forestry Camp. I remember my first night at that camp and waiting for my FN teammates to show up. After driving down a lonely stretch of forest road, I pushed open the lodge's massive wooden doors and shined my flashlight into its cavernous depths searching for a light switch. I was the first to arrive. My flashlight beam swept across a stone fireplace and then across a room full of shining eyes. Suddenly, the beam settled on the face of a snarling bobcat. I actually shrieked before realizing that the animal was stuffed. Thoroughly spooked, I retreated to the woods and set up a tent.

Waiting that night for Tom and Art to turn on the lights turned out to be prophetic. During their course, I learned how profoundly in the dark I was about natural history. Ten months into the FN Program, my knowledge still lacked critical context. Above all, like many naturalists, I was walking over places never paying attention to what was under my feet. When it came to the layer cake, I was wading around in the frosting. I was about to learn that this stubborn allegiance to the life sciences is as limited and outmoded as that stuffed, moth-eaten bobcat. By poring over geology maps, digging soil pits as wide and deep as a grave, coring trees, identifying plant communities, and scouring the land for histories written in stone and barbed wire, the course we called *Siccama and Johnson*

opened up a wide, new world. What I knew about the natural world came alive and my natural curiosity suddenly had room to roam.

At the end of this life-changing adventure, Tom Siccama scrawled across the bottom of my final paper, "If you learned this much in three weeks, you should be teaching this stuff." I guess I took him literally. I spent the next 30 years teaching the layer cake approach to Field Naturalists, and now, with each passing season, to new crops of Vermont Master Naturalists—this bedrock-to-birds exploration of the land around us. It is my bread and butter and it never grows stale.

Siccama and Johnson are not alone in modeling integrated approaches to place-based ecology. Another master of integration is forest ecologist Tom Wessels. In *Reading the Forested Landscape*,[1] Tom shows readers how to tell a compelling story by focusing on the processes like fire, wind, or beavers that each creates recognizable patterns. *Reading the Forested Landscape* is required reading for both the FN and VMN programs.

A third approach that centers on describing natural communities also captures the relationships between organisms, their physical environments, and the processes that affect them. Elizabeth Thompson clearly describes the holistic approach to natural communities in *Wetland, Woodland, Wildland*.[2] Thompson co-taught a UVM course called Field Botany for Natural Resource Professionals for over two decades with Plant Biology Senior Lecturer Cathy Paris, and many talented professional botanists got their start in this class. Reading the forested landscape; pieces, patterns, and processes; recognizing natural communities; and the layer cake are crucial frameworks we teach in the FN and VMN programs.

These frameworks can be used directly in strategies for saving biodiversity. From a conservation perspective, generic terms like "open space" hinder our efforts. I have attended meetings and heard statements like "we have ample open space in our town." I have never heard anyone say, "we have too many Limestone Bluff Cedar Pine Forests." Seeing and describing the diversity of life gives weight to protecting it.

Roving band of learners

Saving biodiversity is discouraging work. Too many times it involves losing a beloved patch of woods, wetlands, or forest anyway. It is impossible work to do alone. Naturalists need to do this work in community. And so will you. Building community starts early in both the FN and VMN programs.

Candidates enter these programs with backgrounds ranging from geology to wildlife biology to botany to English. As teachers and guides, we enlist the students as co-teachers. We might assign each student the task of explaining a different species of tree or bog plant or a wildflower (having each one teach one) so everyone discovers something profound and new. Encourage students to feel *safe* sharing what they know, and you'll be surprised at their newfound fluency. When people gain the confidence that their voices are valued, the learning grows exponentially. Numerous studies that have shown that people retain information better when they are taught by peers than when they learn the exact same material from an expert. And to paraphrase educator Alice Keeler, the one who is doing the work is the one who is learning. By creating a situation where people guide each other through the landscape, the atmosphere changes from the familiar "drag and brag" of the world-renowned authority to a complete, immersive engagement.

Unfortunately, all aspiring naturalists have experienced the "drag and brag." During my early days of birding around those Texas sewer ponds, the lead birder was out front leading and doing most of the spotting and talking—the discussion thus remained exclusively on birds. The acolyte birders walked in the middle and contributed what they could while biding their time to be a leader. The beginners were left in the back, mostly silent, and often struggling to hear and be part of the conversation. By contrast, when a field walk is conducted with more openness, and anything of interest is fair game—including those things *outside the realm* of the designated expert, the interactions become more fluid and collective. Everyone becomes both beginner and expert. The hike becomes more inclusive and the experience becomes shared.

In the FN and VMN programs, we also share food with one another, stopping at bakeries, baking birthday cakes, breaking chocolate bars together, and bringing along a thermos of hot cider on those cold fall days. We encourage humor and play. An early observer of the Field Naturalists said, "You people have a knack for seeing and remembering the ridiculous." For me there is no higher complement. I can remember one day laughing so hard I almost collapsed on the forest floor after a group of us tried in increasingly ridiculous ways to light club-moss spores on fire after reading that they were used early on in photography as flash powder. And really, who knows why this was funny? It was more of a testament to the spark (or in this case, the lack thereof!) made by communal connection and curiosity. Luckily, teaching outside, in direct contact with the beauty of nature, in and of itself generates the energy necessary for learning—through physical movement, through the richness of information flooding our senses, through the sheer density of detail and mystery. Teaching outside can push students beyond abstraction into new mind-sets, simply by their encountering, with all their senses, something much bigger than themselves. If you want to conserve biodiversity, find the people who share your passion and work with them. And don't forget to go outside together.

Telling a story

Hub Vogelmann knew that communication would be a key part of the Field Naturalist training. He wanted people who understood science to communicate with people who make policy. There is a remarkable diversity of ways to tell a story. In the Field Naturalist and Vermont Master Naturalist programs, we encourage the use of many forms of communication in service of better learning about the landscape and sharing our understanding with others. A more integrative way of reading the landscape naturally leads to a more holistic approach to telling stories. And telling stories is both a great way to teach and a key to conserving biodiversity. Poet W. S. Merwin calls this eloquence about the wild world "a forgotten language." Examples of eloquent naturalists are everywhere.

Poems like Robert Frost's "Directive" can be used as a map to walk through Vermont forests. John Elder took this approach to the poem in his book *Reading the Mountains of Home*. Following John's lead, I often use "Directive" to guide people through the settlement history of Little River State Park, where the ruins of farms dot the hillsides. Frost's eye for landscapes in this poem is uncanny; he describes what's found there as "detail, burned, dissolved, and broken off/like graveyard marble sculpture in the weather," a cellar hole "closing like a dent in dough," the rutted forest roads, and the "pecker-fretted apple trees" (this is especially salient, as sapsuckers blaze apple trees with their grid of holes, preferring them to most other trees).[3]

Naturalists can also "write" their own group poems to describe and celebrate a place and foster a deeper sense of community. Upon arriving at a site, each person jots down a first impression on a scrap of paper. And then someone puts them into an order to create a poem. Here are the last two stanzas from a longer group "poem" Field Naturalists wrote about a bog.

> Beauty in minutiae,
> Crows talking in the distance.
> Tamarack needles glowing in sunlight
> Threads of moss rising up from the muck . . .
> Looks like a place we should see mammoths and musk ox—
> Silent, unmoving, timeless.

Bogs seem to bring out poetic language without prompting. Here's another excerpt from a site review at Belvidere Bog written by Claire Dacey, FN '01:

> On a late autumn day, the bog is still—save for the tufted heads of cottongrass bobbing against the backdrop of black spruce. The bleached trunks of dead trees list like the masts of sinking ships in a sea of sphagnum. The living mat is slowly rising. Black spruce survives here because its extremely shallow root system can stay above the waterlogged, anoxic peat layers. It can sprout roots from its trunk, allowing it to stay abreast of the rising peat.

We also make extensive use of what we call "event maps;" these are a free-form approach to mapping a landscape that incorporates drawing and writing in real-time while you're out in the woods. You end up with a not-to-scale "treasure map" of your outing. It's an excellent way to *slow down* while you're outside and distill what's important to you, to record what you're noticing. In Hannah Hinchman's *A Trail Through Leaves: The Journal as a Path to Place*, she introduces event mapping with these words: "It is a simple mixture of words, images and symbols on a page, but it achieves things that drawing alone, or writing alone, seem to fall short of."[4] No surprise then that the mapping of the landscape through choice events and observations dates back to the earliest of human times; for example, Aboriginal tribes in Australia still sing their way across native lands using songlines. Our event maps aspire to be like these songlines drawn onto paper.

Vermont Master Naturalist Larry Montague showed me how he engages students by teaching kids to rap about wildlife, ecosystems, and climate change as his volunteer project. He recruited other Vermont Master Naturalists to his cause, including a school teacher, a tracker, and another naturalist educator. The final performance in front of the whole school was one of the most enchanting things I have ever witnessed. Larry led the whole school in a back-and-forth chant about climate change, and then the rappers took over, performing their parts to the beat of their raps, and ending with their heartfelt refrain:

> These predators and their prey
> It's time for them to have their say
> With climate change we'll lose our game
> The forests we know won't stay the same
> Climate change is ruining the nation
> We can no longer be patient
> These animals, people, trees, and land
> We'll lose it all if we don't take a stand

We have discussed engaging the next generation of naturalists in the previous chapter. Creative approaches like this can be magic.

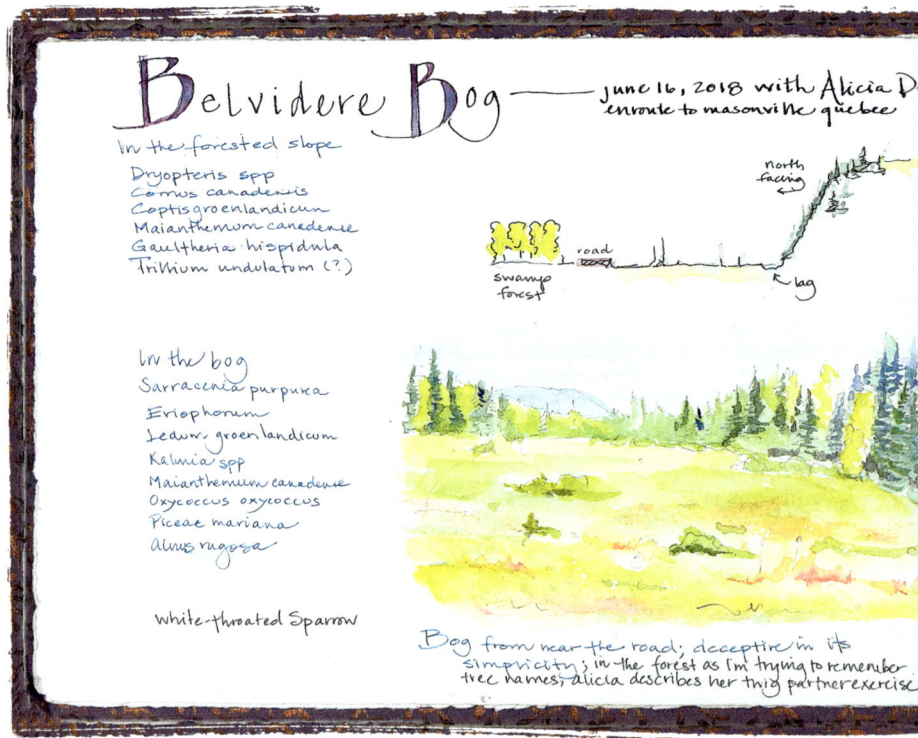

Other FN Program graduates have gone on to become professional science communicators in a range of ways. Thor Hanson is the author of numerous books including *Buzz: The Nature and Necessity of Bees* and *Hurricane Lizards and Plastic Squid: The Fraught and Fascinating Biology of Climate Change*. Rosemary Mosco is a science communicator living in Boston and is the creator of the "Bird and Moon" comics series to tell complex ecological stories.

Are we still saving the world?

The Field Naturalist Program and the Vermont Master Naturalist Program also prepare people to take action to protect biodiversity with real-world projects that are built into the curriculum and the training. Many Field Naturalists use their field experiences working in wildlands around the country to go to jobs in organizations like The Nature Conservancy, Fish and Wildlife, Forests and Parks, land trusts, nature centers, and many more. Vermont Master Naturalists work on projects in their towns.

Lyn Baldwin sketched the beauty of Belvidere Bog on a visit with the author in the summer of 2018.

Together with other Mad River Valley partners and under the guidance of Ashley Madea and Curt Lindberg, the VMN team planted over 70 trees and shrubs in 2020 and also removed invasive species and cared for previously planted stems. Their collaborations along with other groups over the years in this floodplain restoration project earned the 2021 Tree Steward Award from the Vermont Urban & Community Forestry Council Arbor Day.

Giving back

Many Vermont Master Naturalists stay engaged in conservation in their home towns beyond the volunteer hours required for certification. They join conservation commissions, serve on natural resource committees, lead public walks, monitor wildlife, and organize events for planting native plants in public spaces and private yards. Whether or not you are a naturalist, planting native plants is one way to increase the biodiversity of a managed landscape. These rewilding efforts, inspired by E. O. Wilson and Doug Tallamy, encourage people to dedicate public and private spaces to native plants so that the insects that co-evolved with a host plant can lay eggs and their caterpillars will have the food they need to survive. As Tallamy points out, "if you want birds in your yard, you should grow insects."[5] This approach to stitching wild landscapes together by returning native plants to public and private open spaces has gained significant traction in Burlington under the initiative called "Grow Wild," which has a steering committee where half the positions are held by Vermont Master Naturalists. There are many practical ways to save the world by saving the life that shares this planet with us. Planting native plants in your yards and gardens is a radical first step. And you can also join your local conservation or town planning commission. Ask questions on behalf of the wild. Where are the wildlife corridors? What natural communities do we need to conserve? Where are our rare plants? What are we losing through our planning decisions? How are we conserving our wildlands for climate resilience? Keep asking questions. Then boldly suggest hiring ecologists and naturalists to help answer them with you!

Hiking boots on the ground

While dedicated, intentional steps toward conserving biodiversity cannot be overrated, one of the clearest personal conservation memories I have was pure chance. As a rule, I always visit Bristol Cliffs Wilderness Area with the Field Naturalists in October to explore a rare Cold Air Talus Woodland. This natural community occurs in a small band at the base of the talus on private land and features black spruce, labrador tea, and sphagnum mosses. The woodland has a tiny and enchanted elfin quality tucked in next to a full-grown coniferous forest. In October, the fall foliage seen from the open talus is so beautiful. And the rocks are less likely to be icy and slippery than late in the season. But in 1999, I led the FN trip there on a cold November day. We hiked in through an active logging operation. The land to the south had been sold. We passed a skidder and a slew of cut oaks.

When we reached the plateau with the Cold Air Talus Woodland all the red spruce on it had blue blazes. They were marked to be cut. I finished the field trip that day in a fog of distraction and dread. We talked to the logging crew as we were leaving and got the forester's name. I stopped by John Elder's house on the way home and fought back tears as I shared my concern. Luckily, it turned out that the forester's wife had been a student of John's. He helped me arrange a meeting with him the following Monday.

That Monday dawned with a cold November rain. I met the forester and we hiked up to the ledge. The talus was fogged in. No spectacular views. No inverted natural communities in sight. Just the odd little Cold Air Talus Community, which the forester, who knows the woods, could see was out of place. And, I told him, quite rare. There are only four Cold Air Talus Woodlands mapped in the state of Vermont.

Now, I loved the Cold Air Talus, but I also love these big old hemlocks lower down slope that surrounded the ledge. I asked him to please leave the hemlocks as a windbreak. I didn't have time to collect any data, but it seemed reasonable to me that they blocked the wind and the western sun, keeping the site cooler. Sometimes there isn't time for science. All you have time for is a story. Today venerable hemlock stands as a guardian over the Cold Air Talus Community because he agreed with me. He generously donated the trees and John and I raised $200.00 to pay the crew for lost time.

Sometimes timing is everything. That Friday when I brought my class here was the only window of time to make a difference for this Cold Air Talus Woodland. The Friday before, the trees weren't blazed and we wouldn't have known anything was afoot. By the next Friday, the trees would have been cut. We would have walked up on that ledge into a forest of stumps.

How did I come to be there on that particular Friday? The Cold Air Talus Woodland needed someone who could speak to its rarity and beauty.

Cold Air Talus Woodland

Someone who understood what was at stake and cared enough to act. It still gives me chills to think of the possibility that a larger force was at work here. Is it possible to have such an intimate relationship with a place that you know when it is threatened? All I know is that that day I was given the chance to be a voice for this small, wild forest. Sometimes being there is enough. Being there and having a story to tell. Sending naturalists out into the world prepared to tell compelling stories—*on the spot, ready or not*—is key to conserving biodiversity. And there is no question that training the next generation of naturalists to be champions of biodiversity is vital work. Naturalists alone aren't enough. If you care, we need you out there, too.

And for those of you still wondering, all the light switches at the old Yale School of Forestry lodge were in the basement stairwell.

REFERENCES

1. Wessels, T. (1997). *Reading the forested landscape: A natural history of New England.* The Countryman Press.
2. Thompson, E. S. (2019). *Wetland, woodland, wildland: A guide to the natural communities of Vermont* (2nd ed.). Chelsea Green Publishing.
3. Frost, R. (1979). *The Poetry of Robert Frost.* (E. C. Lathem, Ed.) St. Martin's Griffin, pp. 377–379.
4. Hinchman, H. (1999) *A trail through leaves: The journal as a path to place.* W. W. Norton & Company.
5. Tallamy, D. (2007). *Bringing nature home: How you can sustain wildlife with native plants.* Timber Press.

Surely the rest of life matters. Surely our stewardship is its only hope. We will be wise to listen carefully to the heart, then act with rational intention and all the tools we can gather and bring to bear.

EDWARD O. WILSON

Half-Earth Vermont

An Origin Story

STEVEN SHEPARD

There is a comforting elegance to the deliberate acquisition of knowledge for the sole purpose of understanding, comprehending, being enlightened. Knowledge comes in many forms, from many points on nature's compass. We often talk about the "hard sciences" of physics, biology, mathematics, engineering, astronomy, geology, and chemistry; then, from the other side of our brain, across the synaptic chasm, we introduce the humanities—language, literature, philosophy, history, ethics, the arts. The two camps are often regarded as diametric enemies, exclusive binaries, oil and water, two magnetic south poles vigorously repelling each other. But in fact, they are equally important halves of our human whole.

In 1998, Edward O. Wilson, biologist, he of the ants, the quintessence of hard science, reduced the wall between the sciences and the humanities to intellectual rubble with his seminal book, *Consilience: The Unity of Knowledge*. Wilson argues that the quest to unite all human thought, which started during the post-Renaissance period of Enlightenment, must be a hallmark of the twenty-first century. The humanities and social sciences, he says, must join forces with the natural sciences if humanity is to address the increasingly existential challenges that confront us—among them, the loss of biodiversity, and the need to ensure the health of the robust natural world.

In some ways, Wilson is pointing us to what's been learned about healthy ecosystems. When they are rich with diversity, connections, and interactions, and they have space to thrive and experiment, they are remarkably resilient. They face what's thrown at them and continue on their evolutionary, ever-creative journeys. It is this rich ecosystemic approach—in the same way that the power of the humanities and the sciences is augmented when they interact—that leads to the diversity required for a biodiverse planet.

An age of wonder

In *The Age of Wonder: How the Romantic Generation Discovered the Beauty and Terror of Science*, Richard Holmes writes eloquently that a consilience of sorts took place during the Romantic Age, a time of intellectual ecosystems. The scientists of the time made extraordinary leaps of discovery, catalyzed by curiosity and a society hungry to simply *know*. Together, the greatest minds of the time built an ecosystem based on curiosity and a thirst for expanded human knowledge. Charles Babbage, for example, built his Difference Engine, an extraordinary machine with twenty-five thousand moving parts that didn't work until his wife, Ada Lovelace, fiddled with it and gave the world the first functional mechanical computer. She was the only legitimate child of George Gordon, otherwise known as Lord Byron. At the age of 27, she translated an Italian paper on a form of Difference Engine, added to it, messed about with her husband's invention, and in the process created the first programming language—which is why a commonly used language today is called ADA. Lord Byron's close friends included William Herschel who, with his sister Caroline, built the largest telescopes ever created and made remarkable discoveries about the solar system. He was also close with Humphry Davy, who conducted life-threatening experiments about the behavior of gases, advancing human knowledge of chemistry and physiology in the process, including the development of a mining lamp that was explosion-proof and the discovery of nitrous oxide. Davy's best friend was Samuel Taylor Coleridge, the poet who gave us *The Rime of the Ancient Mariner* and "Kubla Khan."

Meanwhile, Michael Faraday discovered crucial things about the behavior of electricity and invented the electric motor. By night, he composed essays and poetry. And physician Mary Shelley was a healer by day, but in the evenings, she wrote her novel, *Frankenstein*. Born Mary Wollstonecraft, she married poet Percy Bysshe Shelley who, while she wrote *Frankenstein*, wrote *Ozymandias* and *Ode to the West Wind*.

The point to focus on is that the scientists who paired themselves with artists, either through love or friendship, were better at their science than they would have been without the inspiration of the other. And the artists professed universally that their abilities to do their own work were amplified and enriched by the perspective of their scientist partners and how they viewed the world.

It is a matter of unlikely allies, this ecosystemic consilience thing. The coming together of disparate, often unrelated schools of thought to enhance meaning, relevance, context, and understanding, is existentially important.

Consilience

I mention *Consilience* in this essay, partly to pay homage to Dr. Wilson, but also to add perspective as I tell the origin story of the Vermont Alliance for Half-Earth. We derive from provenance: a quick online search for long-standing conservation-focused organizations in the Green Mountain State returns 51 entities, and the list does not include national or international groups like The Sierra Club or the World Wildlife Fund or The Nature Conservancy, all of which are active here. These are our organization's partners, our heroes, our champions, elements of the ecosystem. They are, for the most part, homegrown, Vermont-centric groups with a common cause: to realistically conserve the state's wild lands and expand biodiversity for future generations, because it is the right thing to do. And yet, those 51 agencies, associations, businesses, centers, chapters, clubs, conservancies, corps, councils, departments, foundations, institutes, partnerships, societies, teams, and trusts go about the business of conserving the Green Mountains in disparate, yet cooperative ways—a tapestry, a palimpsest, of effort: a consilient effort. And the manifestation of that effort? In the same way that the patchwork quilt of conservation organizations creates collective diverse impact—much more than any one of these organizations could have independently—the goal of *this* organization, *our* organization, is to help people understand that small actions (planting native trees on your land, publishing this book) and new connections (joining with your neighbors and a local conservation commission to protect threatened wildlife corridors), multiplied many times over, can sometimes trigger statewide (or greater) impact. Our strategies attempt to mimic those of healthy ecosystems—tap diversity, build networks, experiment, learn, build on what works. Recall the story of the reintroduction of wolves in Yellowstone

National Park. A coalition of park naturalists, wildlife biologists from the United States and Canada and conservationists worked together to introduce 14 gray wolves into the park in the cold of January 1995. This action triggered a cascade of other changes that over time helped create a more resilient ecosystem in the park and beyond. It also led to a new appreciation of the vital role played by keystone species in natural systems and fueled more reintroduction efforts around the world. E. O. Wilson's half-earth plea isn't to stake off half of the planet and make it forever wild; his plea is simply to cause us to move over and make reasonable and deserved room for the nonhuman inhabitants that we share the planet with in a conscious and deliberate effort of parity and inclusion.

Origins

The American Museum of Natural History is a castle, a fortress, a citadel of knowledge sometimes described as "the planet's field guide." In its unimaginably impressive collection of 34 million items are a 20-million-year-old butterfly, a 21,000-carat blue topaz, 7.5 million gall wasp specimens collected by none other than sex researcher Alfred Kinsey (who else?), and the largest collection of dinosaur skeletons on the planet. The museum sprawls across 2 million square feet of floor space, roughly equal to the shopping area of 20 average Costco stores, or 42 NFL football fields. Even with that much space to work with, though, it can only display a fraction of its holdings at any given time. The museum and all that it houses, curates, and protects is the result of the efforts of hundreds of contributors who share a common goal, captured in the museum's mission statement: "to discover, interpret, and disseminate—through scientific research and education—knowledge about human cultures, the natural world, and the universe."

So, it was only fitting that on October 22, 2018, the museum would host the second Half-Earth Day, a celebration of biologist E. O. Wilson's call to conserve half of the Earth's surface as a way to protect the planet's extraordinary but dwindling biodiversity—including ourselves. It was a landmark event: Wilson himself was there, along with a collection of impressive speakers and an audience of students, teachers, activists, scientists, and ordinary citizens, concerned about the fact that we humans haven't been the best neighbors in the last century or so and it's starting to show.

Among those who attended the event in New York City was a similar mix of people from Vermont committed to the cause championed by Dr.

Our strategies attempt to mimic those of healthy ecosystems—tap diversity, build networks, experiment, learn, build on what works.

Wilson. They listened to the presentations; they enjoyed a conversation about the half-earth movement between Wilson, singer-songwriter-activist Paul Simon, and *New York Times* journalist Tom Friedman; and they had the chance to chat with Dr. Wilson about the idea he captured in his book, *Half-Earth: Our Planet's Fight for Life.*

Jennifer Esser is a librarian and nature advocate who attended the event, and who remembers it as if it only recently happened.

"I had the opportunity to attend the educator workshop that took place in the morning," she recalls, "and Dr. Wilson came and addressed us as a group. That was my first time being in his presence. What a wise, kind man.

"His message was so powerful: There's so much we still don't know about the natural world, and it's so important that we encourage young people to become scientists, and even if they don't go that direction professionally, they should become citizen scientists and learn to observe what's going on around them. For example, I don't have a formal science background, but I've learned to observe and to be curious because it really is a way to connect us to the world and to understand all the ecological interactions that go on around us.

"We also have to understand that Dr. Wilson's vision of half-earth is something of a moonshot. But we have no choice but to try. He's not being an alarmist; he's doing everything he can to inspire us to action, doing the best that we can to protect the natural world that we're part of. After all, if humans are going to survive, we don't really have a choice, do we?"

It's that focus on doing the best we can, the small things, the practical things, the *doable* things, that sparked the Vermont Alliance for Half-Earth to life.

Another participant in the event at the museum was George Schenk. "As a group," he muses, "we share an awareness that nature is in trouble, and that we are dependent on it to be healthy. It's imperative that we do something about that, because it's the right thing to do, and because we have the wherewithal to do it."

The nucleus of the group, the strong attractive force that pulled us together and continues to *hold* us together, is Curt Lindberg. He'll deny it, but Curt is a force of nature, with the ability to cajole, motivate, and inspire those around him.

Sean Beckett is a field biologist and program director at North Branch Nature Center in Montpelier, Vermont. He smiles when he tries to conjure memories of the Vermont Alliance for Half-Earth group's origins. Predictably, his thoughts go to Curt Lindberg as the catalyst that sparked things.

"Here you have somebody like Curt, who is not a biologist. He's an E. O. Wilson fan, but his career has been in the healthcare industry and working with complex systems. But now, because of a passion for biodiversity and the natural world, he has pulled together a group of similarly committed people and has caused things to happen with Vermont conservation that haven't been done before. It just reminds me what happens when we actively go out and find voices and leaders that are not within the traditional, concentric rings of similar backgrounds, but who want to be involved in the environmental conversation."

Our mission

This idea is not new, nor is it human, in the strictest sense. Recent research has demonstrated conclusively that trees in the forest—indeed, plants of all ilk—"communicate" with one another using the strands of mycorrhizal fungi that envelop their roots and travel extensively under the loam and leaf litter and topsoil, a phenomenon referred to by botanists as the "Wood Wide Web." The trees share pest information with one another via their mycorrhizal partners and may even use the fungal strands to send nutrients to ill or young trees that are struggling among them. In the American southwest, sage plants emit chemical signals to warn neighbors of the presence of pests; willows do the same. These organisms are connected in a vast network. And another term for this network? Ecosystem.

Vermont's forests not only survive, but many thrive; the Vermont Alliance for Half-Earth is committed to ensuring they have everything they need to do so. The group's origin story begins with the trip to New York to attend the event at the American Museum of Natural History. Thanks to the initial connections that resulted from that event—human mycorrhizae, if you will—discoveries were made, and talent was uncovered. Through a connection with the University of Vermont's Field Naturalist Program, a second connection was made with field naturalist Eric Hagen, who with support from the alliance authored *A Vermonter's Guide for Protecting Biodiversity*, along with a series of stories about Vermonters and their relationships with biodiversity. Eric is a storyteller—better yet, he is a prolific collector of stories, all with a common theme—that living with the land is far better than living in spite of the land.

Eric's stories and the biodiversity guide are featured on the Vermont

Alliance for Half-Earth website and in this book because they capture the essence of what the group stands for. They give the group's existence gravitas, provide proof that all efforts to live cooperatively, ecosystemically, with the land are worthwhile, no matter how inconsequential they may seem at the time.

Meanwhile, other connections emerged. Thanks to our partner organization, the North Branch Nature Center (NBNC) in Montpelier, the alliance forged connections with educators who were passionate about bringing biodiversity concepts into the classroom. Together they conducted BioBlitzes and other activities as seamless, year-long elements of their curriculum. Into the standard subjects of math and language and social studies they interwove context in the form of linkages to the natural world—a further extension of the knowledge ecosystem. North Branch and teacher members of the alliance even created a weeklong Vermont Biodiversity Educators Institute, now in its third year. Together, the alliance, NBNC, and the education community extended Dr. Wilson's message from half-earth to half-school, half-town, half-yard, half-Vermont.

It should be clear at this stage that the Vermont Alliance for Half-Earth is actually a network of networks, a cooperative ecosystem of skill, capability, commitment, passion, drive, knowledge, a pinch of wisdom, a tat of hope, and a tad of optimism. From that assemblage came the commitment

to produce a book that would tell the story, a chronicle of possibility. A group of writers came together; a structure emerged; and the Vermont Natural Resources Council and Northeast Wilderness Trust donated funds to kick-start the project.

It is the vast heterogeneity of the alliance, motivated by a common passion, which pulled us together and keeps us together. Our shared ideal is simple, but profound: preserve biodiversity by getting out of its way. Through our collective effort, we can raise awareness about the importance of individual efforts that make a tangible difference. Let's be clear: the road from an event at the American Museum of Natural History in 2018 to the formation of the Vermont Alliance for Half-Earth and the production of this book in 2021 was not planned; it happened, through a series of stages, because of the efforts of a vast web of coconspirators, and because it's the right thing to do.

So, who is this Half-Earth Vermont rabble? An interesting question. We are Walter Poleman, a college professor specializing in all things ecological. We are Sandy Fary, a primary school teacher whose classroom is quite literally the great outdoors. We are Liz Thompson, a conservation planner and the coauthor of *Wetland, Woodland, Wildland: A Guide to the Natural Communities of Vermont*. We are George Schenk, founder of American Flatbread and owner of Lareau Farm. And we are writers, designers, analysts, photographers, activists, travelers, elementary school teachers, and representatives to state government.

One goal of the group is to make people feel hopeful and inspired and to actually help create the behavioral change that nature needs.

"That can be hard to do in a world where you're constantly being told that the problems we face are insurmountable," laments Sean Beckett from North Branch Nature Center. "That whole line of thinking is pervasive, and as a result, it can become a little defeating. The beauty of this group, of this half-earth project, is that we manage to do two things at once. First, we acknowledge that, yes, this problem we're facing is a global one. It's monumental, and it's scary. But it's also going to be solved by local action."

The nature center in Montpelier is a good example of the work done in Vermont on behalf of the natural world. The Vermont Alliance for Half-Earth is closely affiliated with North Branch, a place that is unlike any nature center I've ever come across in terms of the breadth and depth of its programs and its commitment to raise awareness about our collective need to have a deep relationship with the other living things that call this planet home.

"The efforts that we see coming from the Half-Earth Vermont group is one of the only initiatives that we've seen that feels like it's in the wheelhouse of the type of work that we do," says Beckett. "From an ecological perspective, it feels like something that will actually make a difference. The group is practical: they don't go out of their way to advocate for cars with better fuel efficiency; they advocate for programs that encourage the planting of native pollinators or doing things that help a colony of bees.

"Our job, at least as we see it, is to serve as the liaison between ecology, natural history, and conservation efforts at the state level. We think of ourselves as national leaders, or at least regional leaders, helping teachers, for example, figure out how to embed the study of biodiversity into their school schedules. The challenge is that teachers are hit all day long with different initiatives that they're supposed to incorporate into the nonexistent free space in their curriculum. So, we work directly with teachers to support what they're teaching and how they're thinking about using the outdoors as a classroom. And by the way, this is not necessarily about teaching natural history, but teaching *with* nature as an opportunity to teach about other concepts. So, I see our role as connecting the science to the teachers and helping our local teachers incorporate lessons about the need for biodiversity into what they're doing. And the Half-Earth Vermont group is committed to helping us make that happen."

Sandy Fary teaches seventh and eighth grade at Camel's Hump Middle School. She's the teacher I referred to earlier whose classroom is literally outside. She's featured earlier in this book ("High Flyers," page 67).

"I strongly believe," she says, "that students learn best through hands-on, cooperative learning. So, my classroom is indeed outside. By the second day of school, we're out of the school building, immersing ourselves in the natural world. And what does that look like? Well, we do a lot of field trips. One of the first trips we take is to the Winooski River. We're in Richmond, and the river runs right through town. So, I kick off the school year with a watershed unit. I take the students onto the floodplain, and we look at the soils that make it up. We also learn about some of the trees as a way to teach about floodplain dynamics in the river.

"From there, we study a variety of natural communities. I take them all over the state. Within their year with me, the students are out in the middle of Lake Champlain collecting data. We're in vernal pools, looking at species and learning about vernal pool ecology. We do four camping trips in the two-year continuum, and in the spring of their seventh-grade year, we do a three-day ornithology trip to the Northeast Kingdom. We're in canoes or kayaks going down the Clyde River, looking for shorebirds. We hike at Moose Bog, searching for specific species of rare birds. We're in Southern Vermont, doing birding as well.

"Basically, I try to give students a sampling of what an ecologist might do. We look at soils, and we work with plants. I have some students who learn the calls of all the frogs and toads of Vermont, while others learn the songs of fifty Vermont birds."

Sandy's curriculum is a quilt of diverse knowledge—diverse being the operative word here. But there's a higher goal at work. Hers isn't just an effort to see how many birds or frogs or lichens a student can identify at the end of the school year; it's to create an awareness, an understanding, of the entire diverse ecosystem, in all its remarkable glory—and, to instill in the students a bias for environmental action.

"All too often, when people hear the word 'wildlife,' they think of it as referring to what are called 'charismatic megafauna'—bears, birds, bobcats, that kind of thing," says George Schenk. "Truthfully, it starts with soil bacteria, and then works its way up to bobcats and bears. When we take that more expansive view, as Sandy does in her classroom, it gives us an opportunity to realize how we can protect the state's biodiversity in different ways."

A team of teachers collecting invertebrates as part of a stream biodiversity analysis during North Branch Nature Center's Vermont Biodiversity Institute for Educators.

Fary agrees. "My goal with my students is to push the idea that the more they know, the more they're going to appreciate and value biodiversity. And if they have a deep understanding of biodiversity, then they're going to invest in the preservation of our lands and become future stewards of their place, wherever 'their place' is.

"Our Half-Earth Vermont group," she continues, "has a parallel commitment. We all have a passion for the natural world. We're all outdoor enthusiasts. We all have a similar mission in that we want to conserve and protect the environment. And each of us is going about it in a slightly different way. But we all share the same mission in that we want the outcome to be the preservation of our land and our water. We're all working cooperatively to figure out what our place in it is. I, for example, go back to the classroom, or back to my community, and look for ways to move this forward, with young people and then with my neighbors.

"Our goal as a group is to push this idea of half-school, half-yard, half-Vermont, half-town. That's doable; that we can execute. And it starts with the kids."

University of Vermont Professor Walter Poleman feels strongly about this position and works closely with Sandy Fary in her programs.

"In many ways, we're lucky here in Vermont," he says. "The state has an outsized reputation and plenty of influence—it boxes well above its weight because of that reputation. It's a beautiful state that cares about the environment, and it's a place people are proud to be from when they talk to other people.

"That's where this idea that we've been playing around with for a long time actually came from, this idea to take Wilson's advice and teach school kids the importance of conserving half of the land. Kids are far smarter than we give them credit for: they can understand the concept of half the school yard, or half of their yard at home, building all the way up to the watershed level. So, we took the half-earth concept and applied it to half of 'place.'

"The important idea is that we need to have half of our mind-sets dedicated to more than just the human world, at any scale, regardless of where we are. By using the natural world as classroom, by teaching curricula at all levels within the context of biodiversity and ecology, we create an awareness early on, and it makes a difference."

Half-mind-set—what an interesting way to think about the mission of the organization, and to keep in mind as you read the stories in this book.

They are proof of the fact, stipulated by Dr. Wilson himself, that when divergent ideas and approaches are allowed to converge through a common, passion-centric mission, magic happens.

Looking forward

Vermont Alliance for Half-Earth *came* together because of a shared belief that the protection and restoration of biodiversity is the single most important thing that we, as sentient beings on this planet, can devote mental cycles to. But we *stay* together as a cohesive, dedicated cohort because we also believe that thinking about biodiversity isn't enough: the thoughts must spawn action on behalf of every living thing on both halves of E. O. Wilson's Earth. The good news is that as you read this book, you will come to realize that the changes required, while planetary in scope, will happen if we all recognize that preservation of biodiversity is an idea whose time has come, and each do our part as elements of a collective whole. Improved economic decisions can help the planet, but more than money is needed. Cultivating biodiversity will require enacting the message printed on every U.S. dollar: *E Pluribus Unum*. From the many, one.

There is no distinction between the fate of the land and the fate of people. When one is abused, the other suffers.

WENDELL BERRY

PART TWO

A Shared Life

Stories about people in Vermont and their varied relationships with the land and biodiversity, based on interviews conducted in the summer of 2019. Part Two also includes an essay from Andlea Brett, who is featured in one of the stories.

Guests at Nature's Table

Food for People and a Home for Wildlife

ERIC HAGEN

George Schenk is the owner of Lareau Farm and founder of American Flatbread in Waitsfield, Vermont. Around the hearth, George's goal is to create delicious food. But on the farm, his goal is to create habitat for wildlife. "I've started to see all of the different lives around me as my partners in this journey," said George, "and as a partner with them, I have both the responsibility and the opportunity to choose to be constructive to the lives of others, or to be hurtful to the lives of others."

A case in point is a little brook that runs under a fallen tree and into a restored wetland back behind the gardens. George explained the scene, saying, "A lot of creating biodiversity is about creating diversity in the habitat, creating the opportunities for life. For decades and decades, if a tree like that fell on this land it would be removed. And that's not wrong from the farm's point of view, but it's not as good from the wildlife's point of view. And so what we're trying to do here is to say, well, where's the balance? What can I do that would be responsible to the agricultural interests of the farm and its food production values, but also responsible to wildlife?" And sure enough, under the fallen tree was an animal track (see photo to right). Do you see it in the muck? It was made by the front paw of a raccoon, and prints of the hind feet were found just a few inches away.

Back in the gardens, George has created structures for wildlife to use, like extra-tall fence posts and brush piles. The fence posts are constantly used by birds, but just a few hundred feet away in the hay field, where the landscape doesn't have the same three-dimensional structure, there's less bird activity. As it turns out, a bird using those tall fence posts once did George a favor.

"We had a population of rats that had colonized the compost pile," George said. "We had them here for two years, and people said 'Oh, you can't have rats, rats are bad.' And I'm like, 'Well, OK . . .' And actually I would sit out there in the evenings and the rats would come out from underneath the rocks and around the compost pile, and they were absolutely charming. You know, they were these little mammals—they were lovely! One of the things they say about good compost piles is that it's good to aerate them. Well the rats created these aerating tunnels. I made excellent compost that year with these rats. But, it turned out to be a lot of rats and I started thinking, 'This is getting to be a little much, maybe I need some help here,' and next thing I know there was an owl using my tall posts, and he was hunting the rats! And now this year we don't have any rats."

The majority of the habitat creation on the farm is focused on the smallest of organisms: soil invertebrates and microorganisms. George explained this, saying "We put down a lot of organic material: our compost and the composted horse manure. In a lot of ways I think of that as the food for microorganisms and soil invertebrates. But like all lives, those soil organisms need both food and shelter. And so the shelter comes in on these leaf and hay mulches." The mulch keeps the soil moist and invites soil invertebrates to move in and create a tiny, vibrant ecosystem. The invertebrates and moisture give the soil a fine-crumb texture and create an ideal environment for bacteria and other soil microorganisms. George calls

A sign in George's teaching garden.

How do we create food that is a joy to eat, but is also responsible to the health and well-being of the people who eat it, and is responsible to the community, and responsible to our environment?

GEORGE SCHENK

Peek under the mulch and you find a tiny ecosystem, complete with its own predators like this spider.

this biomagnification, where increasing the population of some organisms, like worms and ants, increases the population of other organisms, like bacteria and predators.

"We're trying to manage the environment for the garden biologically," George explained. "And as we create these ideal soil conditions the plants can have essentially everything they need to be healthy and dynamic and vibrant. As a result they are more resilient to pests and drought, and they have a greater nutrient density when we eat them."

The farm has islands of habitat for invertebrates and microorganisms scattered about. George explained his reasoning, saying, "I read once that Yellowstone National Park acts as a kind of biodiversity fountain, that a lot of animals come from there, and then they can populate areas peripheral to the park. This idea, then, is: could an undisturbed area dedicated for soil invertebrates and microorganisms act as a kind of reservoir that could populate the adjoining garden area?"

So George built a simple structure out of dead wood and filled it with brush, then left it alone to see what would happen. He also scattered rotting logs and made a bundle of Japanese knotweed, thinking that wasps and other organisms might like them. "It's a little bit like putting a birdhouse out," said George, "except I put out a house for bugs."

Looking out from behind the restored wetland.

"One of my goals here is to show the smallest, simplest things that anybody can do. You don't need a degree in biology to make up a pile of rocks, but that pile of rocks is going to create habitat for soil invertebrates. And you don't need to be a specialist to mulch your tomato plants. And that's OK."

In the fall of 2018 George had the opportunity to talk to E. O. Wilson, founder of the Half-Earth Project, and he said to Wilson, "I'm doing things to try and increase the biodiversity of the gardens, trying to create habitat for the microorganisms and birds and beneficial insects. And I know it's only one little place . . ." Wilson replied, "That counts too."

Looking back on the exchange, George said, "That has turned out to be where I've decided to hang my hat. There's so much that the world needs that I cannot do. But I can do this, and I think that what's important is that we all do something."

Roots in the Land

Generations of Mentorship

ERIC HAGEN

Jim McCullough, Vermont State Representative, started skipping school in third grade. "We used to have to walk down to the corner to catch the bus," said Jim, "which was a perfect opportunity to just never show up at the bus. Parents didn't know because they saw me leave, and then I'd come back at the end of the day."

Jim started putting down roots in the land during his young days, the same land that he and Lucy would raise children on and devote the rest of their lives to. "One of those days I had been watching the men out in the field, and I'd gotten really hungry. They were picking rocks, getting ready to plant. And I was out in the pasture, at the edge of the forest in the shade, watching them do this and got really hungry and ate my lunch. At some point after that I figured it must be time to show up or my parents would be worried about me. And, of course I probably ate lunch at 10:00 a.m., then wandered into the kitchen at like 1:00 p.m., and I got summarily grabbed and marched to school, to the principal's office. So that's just the start of a long career of not being a good student."

Jim did make it to college, though, and in a turn of events found out that he had a love for teaching. He got an offer to work as a teacher, but he and Lucy struggled with their choice: should they continue making ends meet doing construction and other odd jobs or choose a stable career with benefits? "Through my dislike of school I had discovered that there were good teachers and bad teachers," Jim said, "and that the good ones, obviously, worked their asses off. I knew that there'd be no time to teach and do anything with the property if I was going to be a good teacher, and that was the only kind I wanted to be. So that decision got made."

Jim's family has owned the four hundred acres in Williston since 1873. "It's been in his family for so long that he feels a need to honor the family's tradition of taking care of the land and keeping it together in one piece," Lucy explained. "It's been a responsibility that was handed to him from his parents, and his parent's parents."

Jim and Lucy weren't sure how they were going to make a living and take care of the land, but they decided to be OK with the mystery. Today, they look back on a 40-year career of running the recreation business Catamount Outdoor Family Center on their family land, where people of all ages come to run, mountain bike, walk, and ski.

If you were to look down on the property from above—say, from a turkey vulture's point of view—you would see just how close it is to the development of the Burlington metropolitan area, especially Williston and Essex Junction. The looping streets of the suburbs, places that used to be open farm fields and forest not too long ago, spread outward from the urban centers and reach all the way to the property boundaries of the Catamount Center. Being so close to the urban heart of Chittenden County means that Catamount is accessible to many people who might otherwise find it difficult to spend time in nature.

"The place is managed for timber and for wildlife," said Jim, "but it's also managed for people and their growth. We get to be mentors without even realizing it. We don't have to sign up and go be a mentor at school.

When you consider yourself only an extension of the land you can't help but love it. And you can't help but feel that what it has for its own intrinsic value is given to you if you pay attention. Much the same way as the fortunate ones among us get from our parents their values and strengths and beliefs.

JIM McCULLOUGH

Some little kid sees some old fart running up that hill, or riding their bike up that hill, or zooming down it, and suddenly that person became a mentor and that kid's life got changed." Lucy continued, "And it works the other way, too, for us old folks—seeing those kids out there having so much fun. You gotta get on your bike and get out there, or put on your skis and go, so it works both ways. They're mentors for us."

The influence of the land runs deep. People have told Jim and Lucy that having the land available has saved their lives, being able to come to Catamount and recreate—having the quiet, and the peace. Lucy tells of a community member who struggled with mental illness. "She used to come here at the end of her workday and just collapse. And she said without this

A yellow birch that sprouted on an old stump, which has since rotted away. That's why the tree looks like it's vaulted on its roots.

here for her she probably would have never survived." "She would have killed herself," Jim added. "I mean she came right out and said it." Medical research lends support to harrowing stories like this, and the links between time spent in nature and a decrease in health concerns like depression are strong.

The importance of this land in people's lives also shows up in the realm of love and relationships. "We've got children here whose parents met here, and some of whom even got married here," said Jim, "who were brought here by their parents when they were munchkins."

Lucy and Jim tell these stories with passion and gratitude. Their own lives have been nourished by the land and the community interwoven there. "People don't even realize what the importance is, but they come from their cubicle or their assembly line to Catamount, and everything else washes away," Jim explained. "And it's a place they can put down roots, because they've gotten away from their roots. Live in the city, live in a little apartment, you're an attorney, you're a whatever, you've gotten away from that. The trees and the plants all have roots, and so do we, but we as a people have moved on and forgotten our roots, and then people get them reattached when they get here. They don't even know it, but that's why they feel so good."

Their dedication to the land and its stewardship has grown over the past forty years. "Prior to Lucy and I taking over the decision making on the property, the woodlands had been exploited by what we call 'high-grade loggers,' who just knock on your door on any random day and say 'Boy, you got a lot of money out there in trees, how about I take some?' 'Oh yeah, sounds good,' and off they go. But we did away with that practice and got a forester. We got involved with our county forester originally. Then we acquired the services of Green Leaf Forestry, a consulting forester, and we started managing the woodlots for sustainable timber harvest, wildlife management, *and* recreation."

Because they were woodland owners, Jim and Lucy got invited to join Vermont Coverts, a group that helps Vermont landowners better manage their property for wildlife. Attending a Coverts workshop led them to more education about natural areas and their management for wildlife habitat. Jim and Lucy took on projects like restoring wetlands on the property. "In the fall, that pond can literally have 75 percent of the surface area covered with Canada geese," Jim said. "And it's awesome hearing them come in. Early morning they're down there talking about 'Is it time to take off yet?' Or 'Did you get enough to eat?' Or 'Wake up, dammit, it's time to go,' or who knows."

Bears also visit the property in the fall from across the Winooski River in Jericho, Jim says. "The bear come down, swim the river, and come to Catamount and eat wetland plants, and root up acorns and whatever else. That tells you that the bears have an interest in conserving the place."

Lucy and Jim are known to listen to bears, along with our other non-human cousins, and they eventually decided that the best way to conserve their land was to give up ownership. The decision was complicated. They knew that subdividing the property would have made them the most money, but they didn't want the land developed. Jim and Lucy have three children. They, too, wanted to sort out the future of the property. The children wanted to keep the land open, but understood development would net a significant amount more money. Ultimately, their consensus landed with Lucy and Jim on conservation rather than development. "It's such a beautiful piece of property that we'd hate to see it go into housing for a few privileged," Lucy said. "We want it available for everyone." Jim added, "And now the land survives our graves, to put it rather graphically."

Jim and Lucy decided to retain the house and 40 surrounding acres for their family, and worked with The Trust for Public Land to conserve the rest of the property. Today a conservation easement and the development rights are co-owned by the Vermont Land Trust and the Vermont Housing and Conservation Board, while the land is owned by the town of Williston as a community forest. The conservation easement on the property is an agreement that dictates what sort of activities are allowed on the land into perpetuity, and was written together by the McCulloughs and the Vermont Land Trust.

In the end, conserving the land came at a considerable expense and effort, but Lucy and Jim both agree that it was well worth it to protect what they love. "I will also have to say," Lucy continued, "that since we have conserved the property, people say 'How does it feel now that you no longer own the property?' But when I'm out there it's the same property, it doesn't matter who owns it. And if the town owns it now instead of us, it's still the same land. I'm still enjoying it just as much. There isn't that feeling of loss of ownership because I don't think we ever had that greedy kind of a feeling about it. It's just, it's there, we enjoy it, and we use it. We're out there every day. We love it."

Jim and Lucy's hope for the future is that uncountable children—toddlers and old farts alike—will be able to enjoy the property through uncountable years. "This has been our life's purpose," said Jim, "and now it's going to continue."

A Field That Bloomed

Relearning What's Been Lost

ERIC HAGEN

Andlea Brett feels a gravitational pull to water. Many days after work she'll walk down to the lake, or the Winooski, and just stand and listen. It brings her peace. "There's that whole slogan out there now of 'Water is life,'" said Andlea, "and that's real. It's real for the survival of our living beings, and it's real for spiritual life."

Andlea is a social worker, and has lived near Lake Champlain and the Winooski River her entire life. She works for hospitals helping people get access to healthcare. "That often means helping them get insurance," said Andlea, "hopefully affordable insurance. It seems like a little thing, but when someone's on the phone with you, or you have them in person, it could be the last straw between their medicine and whether they're going to eat, or pay their rent or their mortgage. Today I assisted someone who is legally blind with trying to get disability paperwork done because they can't read what they have in front of them, and they don't have support people in their lives to help them do that."

Andlea said it's important to her in her work that she approaches everyone openly, and without judgement. "Whoever they are," she said, "they are valuable, just because they are a human being." Growing up, Andlea's parents taught her that in their Abenaki culture, no one person is more important than another. "Everyone has value, and each person may have a different role that they play, but that doesn't mean that they're less than anyone else, or more important than anyone else."

Andlea said that she knows life isn't fair, but that doesn't mean that some people are more important than others. She said that if she's able to speak for others who can't speak for themselves, then that's what she wants to do. This passion permeates her whole life. "I am on the Vermont Commission for Native American Affairs, and I'm chair of the Vermont Racial Equity Panel, and that has been important for me through the years. When I was younger I always wanted to effect this sort of policy change, and here I am at this age finally part of being able to do this with others."

As children, Andlea and her sisters spent their days roaming through the woods and fields around Colchester, where they grew up. "We used to take off in the morning, and we'd show up for lunch, maybe we'd show up for dinner, wash our feet, and go to bed. And we were always barefoot. They're not here to speak for themselves, but when I'm talking with my sisters," Andlea said, "what's important to all of us is that time in our childhoods when we had complete trust in our environment. Not only was it a more innocent time, but we had a trust in ourselves that the world, as you get older, takes out of you."

When Andlea was a child, the subdivision on Bloomfield Drive (shown in the picture to the right) was actually a field that bloomed. "This is where I used to come and pick the medicinal plants my mom would use if we had an upset stomach, or a cold," said Andlea. "It's funny, even though I was sick, those are some of my fonder memories of my mom, doing up these different brews if I didn't feel good." When Andlea *was* feeling good, her mom would tell her to pick certain plants, and tell her where to find them and what they looked like. "My mother would say 'Don't take all of it,' and 'You

have to leave some.'" Andlea can't remember many of the names of the plants anymore, but she would pick many different wildflowers. She also would pick wild grapes, raspberries, and blackberries. "My mother always said with the plants that we had to thank them for giving us what they were giving us, for whatever we would be using them for."

Along with a neighbor, Andlea's dad grew a huge garden, and also hunted. Andlea said that it wasn't until she was ten that she went to a big grocery store. "I was raised with my Abenaki heritage to the best that my parents could do at the time. Because, you know, my dad was still aspiring to keep up with the Joneses, and also trying to ignore his Abenaki heritage and fit in with the white world, which was impossible. My grandmother used to say he was a throwback. Through the family you would have people who passed and looked white, and then you'd have someone who would be born with very dark hair, very dark skin. No way. You could tell they were Abenaki, and they called it a throwback. And she, my paternal grandmother, my dad's mom, was discriminated against and called very derogatory terms because of her mixed French and Abenaki heritage."

Throughout her family's history there has been a lot of discrimination and pressure to erase their Abenaki heritage. Andlea's paternal grandmother, her dad's mom, was called very derogatory terms because of her mixed Abenaki and French heritage. Growing up, Andlea had a fourth grade teacher

who drilled her accent out of her. For example, the "R" sound in English is replaced with the "L" sound in Abenaki, and Andlea would speak this way. Her own legal name "Andrea" is pronounced "Andlea" in Abenaki. Andlea's teacher would ignore her until she spoke without an accent, and told her that people wouldn't take her seriously unless she sounded educated. "It's a mixed blessing," said Andlea, after a few cars rushed by, "because it is important. The world will not take you seriously if you don't have education, and sound educated. But at the same time, it's losing a part of yourself."

When Andlea looks over Bloomfield Drive now, she reflects on the species that were once plentiful. "Well, there are no more wildflowers here. You know what, those plants are gone. Where are they? Where can I find those plants now? For a period of my life I walked away from all of that, because I tried to fit into the white world. I'm still trying to fit in. But I totally walked away from everything. And in the process, this happened," she says, referring to the now developed subdivision. For thousands of years the land here was full of plants, animals, people, and their intertwining stories. Stories are still being made here, but something is different. "Now it's manicured lawns," said Andlea. "What story is that? Nothing interesting."

The plants here are no longer an expression of the land. They don't tell us what's in the soil, or how their ancestors found a way to live and reproduce in this place. The landscape we see now only tells us which plants respond well to a lawnmower, or what's fashionable in the landscaping industry. All of the old stories have been cut off and scraped away. The land has lost its voice.

"I'm now getting to an age where this field is important," said Andlea. "And now it's not here to pass on to anybody. It's not here to pass on to my kids, you know? It's also sad because I spent that time ignoring my heritage, and now I can't do anything about it. Like the field being gone—cut off. It was too painful to straddle both worlds, so I went totally with the white world, and ignored everything with the Abenaki. I'm wiser now, and more confident now, where I want to take that back, and it's gone. I have to reteach myself." Along with cultural knowledge, Andlea is also trying to relearn the Abenaki language, because in the process of growing up she lost it.

A lot of the land in Colchester has been lost to development, and a lot of Abenaki culture has also been lost to the pressures of colonization, but there are still precious pieces that are left. "You know, my family has used this land for its food my whole life, and my family still uses this land for its food, as we have for generations." Andlea's son Ian provides food for the family by hunting and fishing. "He gets immense pride out of that," said Andlea, "and the fact that our fish come from our lake, not the grocery store. And at the same time, he is very solemn about the fact that he took that fish's life, or that bird's life." Before cleaning an animal Andlea's family says a prayer. "You thank the animal for giving its life, so that you can have life." But like with our landscapes, the pain of loss is still near. When Ian was growing up, Andlea's father was preparing to teach him the words in Abenaki, but he passed away before he could. And so they say their prayers in English.

Walking along Colchester Bog, her childhood stomping grounds, Andlea continued to speak about the land today, and her family's relationship to it. "I still believe, and was raised believing, that all these living beings are our relatives—the trees, the plants, the animals." This belief came from Andlea's Abenaki heritage, but conventional scientists know this too. Life began on Earth at least three billion years ago, and from those first simple organisms, all life, including us, arose. From an evolutionary perspective, plants are our cousins, and the other animals are our brothers and sisters.

"We have an eagle that lives near us," said Andlea. "So when we clean the fish and bring those remains back down to the water, we say a prayer of thanks. We give back to the land and know that other fish, or that eagle—they're going to come eat those remains." In this way, there is no lasting death in nature, because after death, the bodies of all organisms turn again into new life. It's in landfills and under pavement that death remains permanent.

"What if this were gone," asked Andlea, motioning to the bog, "and future generations have to read about it in a textbook, and only know about it because they see it in a picture? That would be heartbreaking. I think about what's already gone, and what I only see in textbooks, or in pictures. And that makes me sad."

Walking down the road next to Colchester Bog, Andlea continued, "When we lose nature, lose our connection to the land, how does that change society? How does that change us? I think we lose a part of our humanity. People, I feel, start thinking that they are in total control of everything, and they forget that, no, we're not in control. We are part of the land, but we do not control it. We have to take care of it."

"We are all connected," Andlea continued. "We are. And we might think we're making a decision that's not hurting anyone else but ourselves, but that's never the case. Never. It's never just a decision we're making for ourselves. Some Native Americans have always thought, how do we preserve this up to seven generations? And if we're teaching our children seven generations, and they teach their children, then maybe we can preserve life on this land."

Andlea said she doesn't know how we shift perspectives to see like this, to see the world as a place that needs care in order for life to continue to be abundant. "I don't know how we effectively educate people to think beyond this one moment in their lives. But we aren't going to be here forever. What's beyond our one moment? What about other people's moment, or their children's moment?"

"I have more empathy, and understanding, and allowance," said Andlea, "for that person who's just trying to eke out a living, than the person who has the means, and should have the brain, and the time, to be able to think differently and act differently. When folks are too poor, and all they're worrying about is whether they have enough food, they're not going to care about biodiversity. They're going to care about whether they can feed themselves and their families. The people I work with want to be able to care about the environment, at least that's what they've told me, but they don't have the luxury of caring for the environment. And then, if you're rich enough, sometimes the attitude I've seen has been that they don't have to follow the rules."

And with at least one major metric of environmental destruction, Andlea's observations are accurate. The world's richest 10 percent of people produce half of the global carbon emissions, while the poorest half of the planet produces just 10 percent. And if you look at the producers of goods instead of the consumers of goods, only one hundred companies are responsible for 71 percent of global greenhouse gas emissions. "The middle class has been convinced that somehow we're taking care of the really poor and the really rich," said Andlea. But when you look at the numbers, you see that that doesn't make sense.

It can't be the only solution, but governments from the wealthiest nations have to shoulder the majority of the responsibility for reversing climate change and biodiversity loss. For example, with only 4 percent of the global population, the United States is responsible for 25 percent of all historical carbon dioxide emissions.[1] To make our governments act we need to pressure politicians to take these crises seriously, because so far they are failing. At the time of this writing, the most recent United Nations climate summit, COP26, ended without any agreement to limit global warming to acceptable evels.

Climate change is projected to be the largest cause of biodiversity loss globally by 2070.[2] One reason for this is the warming of habitats that animals and plants live in. As the climate warms, many populations of plants

and animals will have to move north to cooler locations, but if their paths to suitable habitats are blocked by human development, they will die.

The Intergovernmental Platform on Biodiversity and Ecosystem Services (IPBES) chairman Robert Watson described the interconnection of our environmental crises like this: "Land degradation, biodiversity loss, and climate change are three different faces of the same central challenge: the increasingly dangerous impact of our choices on the health of our natural environment. We cannot afford to tackle any one of these three threats in isolation—they each deserve the highest policy priority and must be addressed together."[3] This is partly because keeping ecosystems intact is one of the best ways of keeping carbon out of the atmosphere, and restoring ecosystems is one of the best ways to sequester carbon.

"I still like to think there's hope," Andlea said, despite all of the challenges she sees. "We had an Abenaki revitalization week up at Northwoods Stewardship Center, in I think it was June of 2017," Andlea said, looking at milkweed pods along the side of the road. "It was a week-long camp, and they brought in teachers for relearning the medicinal and edible plants. Basically getting back to those basics. All of us who took part in that initial camp, our lives have changed in ways that I don't think any of us could have foreseen. It was an amazing week. It's what helped me be able to come back, and be like 'Yes, this is my role in this world, as an Abenaki.' And as much as I can, I'm getting back to the plants that I can use to treat myself, and heal myself. I've also been relearning the language, because as I said to Jesse Bruchac, our language teacher, a people without their language have no civil rights. A people without their language are dead. We lose our culture. We lose ourselves. And so I'm trying." Andlea said that as she relearns the language she can't help but reconnect with her environment. This is because in the Abenaki language, things in nature are grammatically imbued with life. Plants and animals, and even water, the wind, and stones, are all spoken about with a sense of animacy and agency. Things like buildings and streets, on the other hand, are treated as lifeless objects.

In her book *Braiding Sweetgrass*, the botanist and author Robin Wall Kimmerer writes of her ancestral language, one that is closely related to Abenaki, "So it is that in Potawatomi and most other indigenous languages, we use the same words to address the living world as we use for our family. Because they are our family."[4]

For the rest of us who aren't Abenaki, what language do we learn? We have to learn the language of the land—listen to what the water, the plants, and the animals are saying. Understand what they need to thrive, and let their stories continue on. Because if we don't learn that language, if we don't learn how to give the natural world the treatment it needs to be healthy, then the land will lose species, relationships, and stories that can't be replaced. We can't allow that to happen.

"Our environment has given and given and given," said Andlea, "and are we giving anything back?"

REFERENCES

1. Hickel, J. (2020). Quantifying national responsibility for climate breakdown: an equality-based attribution approach for carbon dioxide emissions in excess of the planetary boundary. *The Lancet Planetary Health*, 4(9), e399-e404.

2. Newbold, T. (2018). Future effects of climate and land-use change on terrestrial vertebrate community diversity under different scenarios. *Proceedings of the Royal Society B: Biological Sciences*, 285(1881).

3. IPBES. (2018). *Media Release: Worsening Worldwide Land Degradation Now 'Critical', Undermining Well-Being of 3.2 Billion People*. IPBES secretariat. https://ipbes.net/news/media-release-worsening-worldwide-land-degradation-now-%E2%80%98critical%E2%80%99-undermining-well-being-32.

4. Kimmerer, R. W. (2013). *Braiding sweetgrass: Indigenous wisdom, scientific knowledge and the teachings of plants*. Milkweed Editions, p. 55.

A Field That Bloomed

The Winding Path of Reconciliation

ANDLEA BRETT

> *So this is what we truly believe. This is what reinforces our spiritualities: that no being is greater than the next, that we are part and parcel, we are equal, and that each one of us has a responsibility to the balance of the system.*
>
> **ALBERT MARSHALL, MI'KMAQ ELDER**

When Eric contacted me in the spring of 2019 about doing an interview together I almost canceled on him and "A Field That Bloomed" was almost never written. Because of the history of how the Abenaki have been treated in Vermont, I wasn't sure if I could trust this person I had never met before. It's something I believe many of my fellow Abenaki feel the same about. You must prove yourself to us first, and trust is hard earned. And so now, again, I am taking a chance with all of you total strangers who are reading this, trusting in our shared desire to take care of the Earth that has cared for us.

I grew up in a time in Vermont where hiding your Abenaki identity was a matter of survival. I remember my paternal grandmother begging me and my cousins to "pass" as white. "Just pass," she would implore us, "because if eugenics happened once, it could happen again." She herself was harrassed, called a "half-breed" for being part French and part Abenaki, and she lived to see many Abenaki sterilized under Vermont's eugenics laws. The last sterilization in Vermont happened in 1957, just a little over a decade before I was born. But though my parents and I were spared the direct effects of eugenics, its legacy and the culture that supported genocide were ever-present around me for much of my life. I grew up hearing the story of a friend's home being burned to get him off his land, and as a child I was asked to watch and warn the adults if the police were coming to stop our gatherings where we would dance and drum. This all means that when I was young and my grandmother begged my cousins and me to "pass," we did our best to do just that.

But things were slowly changing, and during my teens and early twenties Vermont went through a small culture shift and it felt slightly safer to "out" yourself, so to speak. My cousins and I started to acknowledge our heritage by saying we were "part Abenaki," but this partial recognition was also a reflection of my own inner confusion and sadness. I didn't fit in with the white world, but there was also no way to be safely and fully Abenaki. As a child, I was picked on and picked apart just for being "part Abenaki," and in self-hatred tried my best to ignore the Abenaki part of myself. In this tumultuous period of growing up I forgot the language I was born into along with much of what my parents and community had taught me. I was in pain, but there were literally no surviving Abenaki spiritual leaders left to help me navigate the healing I needed.

This lack of guidance changed for me one day in the '90s when, in college as an adult learner with a young child, I met a couple of Mi'kmaq elders who were able to help me. After sharing my story, they told me, "There is no part of anything, you are a whole human being. You are Abenaki." This was the first time in my life anyone encouraged me to fully embrace my identity. It was the beginning of a long-needed healing process, and I continued to work with these elders for four years.

Though I started this personal work in the '90s, I still avoided public engagement for many years because of how I had been treated for being Abenaki. Then, in 2017, my brother, who we had lost tragically and suddenly, came to me in a dream and told me there was an opportunity coming up, and I had to promise to take it. Not long after, a coworker told me about the Abenaki Revitalization Week, and my boss—having overheard our conversation—told me I had the week off work. I had to go.

It was a week of no electronics, no phones. Back to nature with learning traditional fire-starting skills, making shelter outside and sleeping in what we had made. I started the relearning of both edible and medicinal plants, as well as my Abenaki language, parts of my heritage that had mostly been lost over the decades and centuries of harassment.

I am forever grateful that my brother sent me to that week of communal Abenaki revitalization. It helped heal me, bring closure from my brother's death, and open doors I never thought possible. I found the courage and support to take on public roles advocating for the Abenaki and broader racial justice. I have now served on the Vermont Commission for Native American Affairs and chair the Governor's Racial Equity Advisory Panel. I am also proud to be working on Vermont's new process of truth, accountability, and reconciliation regarding the historic treatment of Indigenous people in our state.

Reconnection with the Abenaki community during Revitalization Week also led to an interview with Eric, which turned into "A Field That Bloomed." Recently, Eric asked me what parts of the story most resonate now. That story tells of a lot of pain and grief that I've gone through in my life. Fortunately, what most resonates with me now is not the confusion of being an unwanted person, but the trust in myself, my environment, and my community that I had as a child. I look back and remember the confidence and comfort in my own skin from those times of drumming, dancing, gatherings, picking medicinal plants for my mom, and running free outside. I have finally been able to reclaim this state of being for myself. For the first time since my childhood I am wholly comfortable in my own skin. Now I

am excited for new gatherings of Abenaki, the continued relearning of our language, dancing, drumming, and that comfort in my own skin, so to speak, that comes with being part of a community of people who not only understand me, but accept me without ridicule.

It's important for me to share this personal story because many of the Abenaki, as well as other Indigenous peoples, are working hard to heal ourselves from generations of trauma and persecution by reclaiming our heritage. My hope is to build the vitality of our communities and cultures so that more of us can heal and become whole. These efforts might seem strange or imperfect (whose efforts ever are perfect?), but the support of our Vermont community is needed nevertheless. Just as the land did not ask to be colonized, divided, and conquered, neither did we. Like the forests of Vermont, we have been here since the glaciers retreated. Because we are all connected, and all a community, I ask that in all of your efforts to conserve biodiversity—at home and abroad—you ask yourself how Indigenous peoples are affected, and support us even when conservation is not the issue that brings us to the table.

What's important to me now is ensuring that I can pass my Abenaki heritage and a healthy Earth to my children. My hope is that the guidance in this book will help us do that. We need to make sure that there are clean, healthy waters that we can fish from, healthy lands that we can hunt from, and healthy populations of native plants that we can eat and make medicines from. We need to sustain our ability to feed and take care of ourselves not only now, but also for the next seven generations to be able to do the same things. If each coming generation thinks and acts the same, and if we remember that in all our actions we bear responsibilities to each other, we will be able to preserve this Earth that has sustained us.

The Abenaki are the people of the Dawnland. Let's pray the sun up each day as it rises in the East. I will be smudging and sending prayers to the Creator for our Earth to be healed, along with myself and the hearts and minds of my fellow Abenaki and Indigenous peoples. I will pray that we all remember that we cannot survive as splintered parts. We are wholly human and capable of caring for and sustaining the Earth that has cared for us.

I sincerely thank you for joining me in my story. I appreciate your interest and heart in wanting to sustain our Earth together, as it has sustained us through much abuse. Please be respectful and responsible with each step you take and decision you make, not as pieces and parts, but wholly human. K'tzi Wliwni, ta wlipom'kni.

An Unusual Forest

You Can Manage Land and Love It Too

ERIC HAGEN

"When I first bought this property and was laying out the timber harvest, I encountered this young bear. It was in July, and it was one of those males that overwinters with his mom its first winter and then got kicked out the following June. These young bears are like teenagers: they're kind of dumb, foolish, and brave, and also relatively small."

Ethan Tapper was marking trees when he heard a sound. He looked up and saw a young bear immediately take off. Naturally, Ethan followed after it, all the way up a ridge where it started to climb a tree. "I was like, 'Oh, I'll just leave him alone,' kept doing what I was doing. And then I heard another sound. He was back! And he kept on sort of running away, and then coming back to check on me. At some point he decided that I wasn't a threat. I was probably twenty feet away from him, and he just started doing bear stuff. He was just eating leaves, tearing up a rotting log, rolling on his back and rubbing his belly. It was wicked cool."

Ethan Tapper is the Chittenden County Forester, and spends most of his day helping other people manage their land. "One of the gifts of my job," said Ethan, "is I walk on a ton of different lots all over the place, and so I have more of a big-picture understanding in this county of what the forests are like."

When Ethan walks people's land with them, he's reading the landscape, trying to understand how the land and its history shape the forest that's growing today, and how the forest will develop into the future. Understanding the inner workings of the forest lets him help people write management plans that both produce local resources and keep the forest healthy.

So in 2017 when Ethan walked onto a property in Bolton and was considering buying it, he had an intuitive understanding of what most of the forests in the area look like. This place, he knew, was strange. "I walked out here and thought 'this place is weird.' Unusual natural communities. Unusual forest types. All these weird little situations that are just unique and fascinating about it. But also it's just weird. I really like it!"

Ethan spoke with an old co-worker who sells forestland real estate, and the guy called the place garbage. It's got steep slopes and rocky soils, and previous owners "high-graded" the woods repeatedly, removing the most valuable trees and leaving the worst ones to grow bigger and drop seeds. In some places invasive species crowd out native plant seedlings, and the whole property has a deer browse problem, which makes the challenges caused by past logging worse.

The south-facing slopes of the Green Mountains in this part of Vermont support a lot of oak trees, which like the warmer, drier sites that these locations provide. Oaks are valuable timber trees, and are projected to grow better in Vermont as the climate changes. They are also incredibly important for wildlife because they provide vegetation for animals like caterpillars and deer to eat, as well as acorns for animals like mice and bear. By feeding certain animals at different times of the year, which in turn interact with the rest of the environment in various ways, oak trees act as an important component in our local food webs.

The problem for oaks on Ethan Tapper's property is that deer love to

A young red oak whose torn branch tips are a sign of heavy deer browse.

A beech tree severely affected by beech bark disease.

With smooth grey bark, this "clean" beech isn't showing any sign of beech bark disease.

eat their buds and twigs more than most trees. All of the big healthy oaks were cut out by past landowners, and now there are so many deer around that they eat the new oak seedlings before the trees can get tall enough to be safe from their hungry mouths. Walking around Ethan's property, almost all of the young oak trees have torn branch tips made by deer nipping off the tender young shoots. Now that wolves and catamounts are extinct in Vermont, the only significant predator of deer are humans, and so deer are more abundant now than they ever have been.

Another challenge on Ethan's property is beech bark disease, a fungal infection spread by introduced insects from Europe. Beech bark disease causes the bark of beech trees, which should be smooth, to blister and pock. In a last ditch effort before the trees die, they send their energy into shoots that sprout from their roots, filling the understory with young beech saplings. Deer don't like to eat beech nearly as much as other species, and so some parts of Ethan's property are turning into a monoculture of diseased beech.

"I sort of feel," reflected Ethan, "that this property in some ways is like this injured compromised thing that I'm trying to take care of. But everything about this place is hard, you know?" The land is hard to care for, but it's beautiful, and Ethan loves his relationship with the place. Everything Ethan is doing is an act of caring for the land. In the middle of a clearing is a "clean" beech, which means it isn't showing signs of beech bark disease, and its bark is almost completely smooth and unscarred. Before Europeans brought the disease to North America, all of our beech trees were smooth like this one. Many foresters are hoping that cutting the diseased beech and leaving the clean ones will help future generations of the species be more disease resistant. So Ethan is leaving this beech as a "legacy tree," one that he will never cut, letting it live out its life as long as it cares to.

Just like with the beech, Ethan tends to many different timber and non-timber species in the forest. "Something that really makes me happy," Ethan said, "is being able to manage for all this different stuff. I can just be like 'Yes, I love oak, and I also love hazelnut, and I also love serviceberry. I also love red pine, and I will never cut red pine."

Where the forest was most affected by deer browse, beech bark disease, and bad historical logging, Ethan cut out big patches in an effort to increase the number of oak trees on the property, and correct for some of the harmful effects of recent human disturbance. These patch cuts are good for oak because oak have a chance of out-competing other species when there is more sunlight reaching the ground. Bigger patches also give oak seedlings a better chance of escaping the curious muzzles of deer. "In 2017 I bought this property and I realized that it was an oak seed year. And so I immediately put in a bunch of these bigger patches with a whole tree chip crew. These patches were areas that basically had no healthy trees. And within the patches I left every tree of any species that was healthy. I left a lot of oak, with the goal being that those are going to seed into these patches."

Ethan summarized how he approaches forest management like this: "My understanding of how healthy forests grow, and what responsible forest management is, has ultimately boiled down to the development and maintenance of diversity in the forest." Ethan uses the word diversity instead of biodiversity when he's talking to landowners, because in a forest ecosystem both species diversity and structural diversity are important.

Structural diversity means considering a diversity of habitats for all the creatures that live in a forest, from mushrooms and millipedes to barred owls and bears. A forest with a lot of structural diversity has big old trees with lots of holes, and also young sound trees. It has live trees, and dead trees, and rotting logs. It has mature forest and open meadows. From the forest's perspective, a wind-thrown tree isn't an eyesore, but a source of food and shelter for years to come. It also gives the seedlings on the forest floor a window to the sky, and a chance to reach the canopy.

Old-growth forests, the forests that dominated Vermont before European settlement, naturally had all of this diversity. More trees were allowed to grow old and die. Hundreds of years of small fires, windstorms, and floods produced a patchwork of old and young forest. The biodiversity native to Vermont evolved to live in this structurally diverse landscape.

"To me," said Ethan, "that is the way that these forests are sort of supposed to grow, as expressed by themselves. Right? Because in old-growth systems, that's how they grow. And the only reason that most of our forests are not growing that way right now is because we cleared everything for pasture a hundred and fifty years ago."

Old-growth forests also store a lot more carbon than young forests do. Trees turn gaseous carbon dioxide into solid wood, and old forests have both more living wood and dead wood than young forests. More carbon in wood means less carbon in our atmosphere. Today, foresters are learning to mimic the structural diversity of old-growth forests to both store more carbon and provide the habitats needed to support healthy ecosystems.

A patch cut with red pine and oak seed trees.

Ethan Tapper explaining responsible forest management.

A pocket of rich northern hardwood forest growing in a cove on Ethan's property.

Purple coral fungus, wild ginger, and maidenhair fern found in the moist and rich cove.

A rocky ledge covered in reindeer lichen and stunted oaks.

Besides leaving legacy trees, Ethan purposely leaves dead wood in the forest to provide habitat for mushrooms and grubs—which is what that young bear was looking for when Ethan watched him tear into a rotten log. Ethan also leaves the crowns of downed trees intact and messy-looking to provide a natural barrier against deer, so tree seedlings can grow tall before they are browsed to death.

In short, managing for old-growth characteristics is a way to accomplish many of the management goals that are important for biodiversity and climate resilience, part of what Ethan Tapper considers to be "good forest management." "I feel like I could make an argument for good forest management to almost anybody," Ethan said, "because good forest management is often good commercial management." This is because, unlike short-sighted extractive logging, managing for the health of the forest often increases the value of your forest over time.

"One of the challenges of my job is figuring out an argument that works for everybody. And I will talk about totally different things when I'm talking to different people coming from different places. But it will ultimately all access the same themes of diversity."

Beyond just management goals, Ethan also believes in the intrinsic rights of the creatures of the forest, and of the forest itself, to exist and be healthy. But to Ethan, that doesn't mean being totally hands-off. "Increasingly I think about forests as being an organism, which is comprised of many individual organisms, like a system. I've also become very comfortable with death as part of that system—understanding that death is required by forests in order for forests to be healthy. I'm becoming more comfortable with the cutting of trees and the killing of deer as part of the management system, and also the killing of invasive species. It's all complicated."

In line with Ethan's belief in managing the forest for diversity, Ethan leaves some parts of his property entirely alone. Higher up on the property is a cove, a sort of flat basin in the hillside where water and nutrients collect. The richer soils here support species of trees that aren't found in other parts of the property, like sugar maple and white ash. "This is a completely unique site on this property," Ethan said, "and so that's what I think makes it really special. It's not my job to make it productive forest land. It's not my job to make it anything that it's not. It's just my job to respect the diversity that this site provides that's different from other sites on the property."

Here in the cove, new and sometimes strange creatures make their living. Purple coral fungus thrives in the moist soils. So does wild ginger, with its heart-shaped leaves, which only grows in nutrient-rich, shaded soils. It's the same story with maidenhair fern, along with many other species.

Down out of the rich woods, through the open patch cuts, and then through a tangle of beech, Ethan walked out to a rocky ledge covered in reindeer lichen and stunted oak trees. "I try not to come out here that much, because I just sense that it's extremely sensitive. But this is a spot that I found the second time I came here. I was standing up here and I thought to myself, 'Well, I've got to buy this place.' And also deer and bear for some reason like to come up here. I don't know why."

To Ethan, this property is more than just resources that he has legal rights to—it's a place to be in partnership with. "I come out here and I'm like, 'I can't own this. What an absurd idea!' I'm the owner of this piece of land, but do I own this rock? It doesn't make any sense." And yet Ethan continues to cut trees, hunt deer, and remove invasive species. "It's not that I think these places belong to me, or that I'm smarter than them, or better than them in any way. But I just feel a responsibility to take care of all of this, and all of the organisms here."

 Forever Wild

Quiet Refuge and Space to Listen

ERIC HAGEN

"Out here, I'm just a guest . . . the plants, insects, and animals, they live here." Sue Morse was leading me on a walk through one of her wildlife study areas near Bolton, Vermont, and I had just asked her, "What does your relationship with the other organisms on Earth look like?" Sue kept moving, and I trailed along, peppering her with questions as we picked our way through the homes and neighborhoods of countless plants and animals.

Sue is a wildlife ecologist and teacher of natural history. As the name of her organization Keeping Track® suggests, Sue is well known for using tracking as a way of understanding where wildlife live and how they use the landscape. She puts these skills to work, helping to define core habitat for wildlife from "Florida to the Arctic."

The photograph to the right shows Sue in the Arctic islands, where she has been researching wolves, polar bear, musk ox, and caribou for her upcoming book. Sue has spent over 13 years studying these animals in Alaska's Arctic National Wildlife Refuge and throughout the Canadian Barren Lands. Reflecting on the importance of these ecosystems, Sue told me, "My definition of half-earth would be to start with the boreal forest and the Arctic. One third of all the trees on the planet grow in the boreal forest."

Sue has tracked jaguars in Arizona, panthers in Florida, and wolves along the Arctic coast, but she lives in Vermont, and spends most of her time working to understand and protect the wildlife that we share our state with. One of Sue's local efforts, the Chittenden County Uplands Conservation Project, recently crossed a major milestone.

Decades ago, while studying wildlife in the area, Sue had the idea to conserve large tracts of high-quality wildlife habitat in Chittenden County, the most developed county in Vermont. In 1999, that idea turned into a meeting in Sue's living room, which led to a collaboration with local conservation partners like the Vermont Land Trust, the Jericho Underhill Land Trust, the Richmond Land Trust, the Bolton Conservation Commission, The Nature Conservancy, the Green Mountain Club, Keeping Track, and more. With the creation of a town forest by the town of Richmond in 2018, the project has now surpassed more than ten thousand acres of conserved core habitat in Chittenden County. And, as Sue reminded me, they're not done yet! Sue describes this project as "conservation for the commons"—people getting together, determined to protect their local wildlife. "What I like to tell people," said Sue, "is 'You can do it! Don't wait for the next guy!'"

The Chittenden County Uplands Project has helped to conserve large connected blocks of forest that stretch from the peaks of Mount Mansfield and Camels Hump all the way down to the Winooski River. These large swaths of intact forest are important for animals, plants, and other critters because they contain a diversity of habitats. Each of the different habitats is home to different plants and animals, and the fabric of habitats together provides for the full needs of wide-ranging animals like moose, bear, bobcat, and more.

Sue understands the importance of conserving undeveloped forests for recreation and timber, but what really excites her is wilderness—places

Out here, I'm just a guest . . . the plants, insects, and animals, they live here.

SUE MORSE

where animals can live their lives on their own terms, free from the disturbances of heavy human activity.

Some of the land Sue is taking me through today is conserved with a forever-wild easement from Northeast Wilderness Trust, which legally requires land to be left free from logging and mechanized recreation in perpetuity. Sue said she likes that the forest will be able to remain old growth and that it will continue to provide unique functions for the animals that live here.

One of the first stops on our walk is a place Sue calls "Babysitter Swamp" because it's a place where a mother bear that Sue studies takes her cubs when she needs to run errands. Babysitter Swamp contains a perfect combination of features that help mom raise her cubs in peace and security.

Tip-up mound in a sunny wetland, where all kinds of critters and mysteries flourish.

A balsam fir with a strange frayed bite mark.

The teeth of a replica bear skull (named Yorick) fit perfectly into a bite mark found on a paper birch tree.

Swamps are rich sources of food that green up early in the spring and support a diverse abundance of plants and insects throughout the year. Babysitter Swamp is particularly special because it has a high proportion of big, old hemlock trees next to it. Sue has found that bear cubs regularly climb into the topmost branches of these trees to be safe and hidden, from both predators and the hot sun, while mom goes to the "grocery store."

Sue has a lot of memorable names for features in the natural world. For example, she calls openings in the canopy "forest windows" because they let the sunlight into an otherwise dark understory. This light provides an opportunity for new life to tangle itself into existence, and insects, birds, and other animals are drawn into the bounty, one after the other. Wetlands commonly create these forest windows because it's hard for trees to establish themselves, and then hard for them to stay upright once they do. Sue explained the importance of these wetland edges to bobcats in her study area: "Wetland edge, in all seasons, affords bobcats opportunities for prey that are somewhat more concentrated. So that means that the unit of effort expended for hunting that terrain is more apt to be rewarded."

Sue explained that it's not easy being a predator. Imagine trying to chase down a rabbit and catch it with your hands! So, if you've got kittens to feed, it's really important to find habitats that support higher concentrations of prey. "The ecotones at the wetland edge are all great spots for a variety of prey, from snowshoe hare and grouse to turkey. Turkeys will work the edge, too," Sue explained, just like the bobcats do. "They put themselves in a place where there are more open habitats, and therefore a greater abundance of insects, because the poults (baby turkeys) need that extra burst of protein that the insects provide. So, little glades, seeps, and other forested wetlands with any openings is where it's at for them."

I asked Sue why she chose to take me here. "Wetlands," Sue responded, "are alive with all kinds of mysteries and critters that are flourishing here." On the other side of the marshy opening, Sue showed me a balsam fir tree with an unusual wound. Sue explained how bears will sometimes anchor their top canine into a tree and drag their lower canine across the trunk, biting the tree with the effect of mixing their saliva and scent into the frayed wood. This bite leaves a characteristic sign on a tree trunk, a puncture ("dot") where the the top canine anchors, and a frayed drag mark ("dash") where the lower canine gouges. Some of Sue's students have affectionately named this dot-dash pattern "bear Morse code."

Many animals pick conspicuous places in the landscape, like trail crossings and prominent trees in a wetland, to leave scent messages to others of their kind—"exchanging the daily news," as Sue calls it, or alternatively "in-your-face-book." By scraping or spraying, animals might be saying *this place is occupied, please move along*. Or, *I'm looking for a special someone, and now you know where to find me*, or any other number of things.

Sue and I kept walking while she explained these things to me, and soon enough we came across an almost unbelievable demonstration of the importance of quiet refuge and ecotones for biodiversity. Through the trees we could see what seemed to be a couple of great blue herons crowding into one giant nest. But upon further inspection, what looked to be two herons turned out to be eight, spread between three nests in one dead tree! The culprits responsible for this scene of wild abundance? They didn't even bother to hide the evidence. Beavers built this lodge under the rookery tree, they built another lodge (out of the frame of the photo to right), and they built the dams responsible for turning what was once a closed canopy into an open series of ponds and meadows.

Beavers are a keystone species. This means that through their efforts of providing a living for themselves they create conditions that also provide abundance for many other species. Sue has watched this pond ever since the beavers first impounded it in the 1970s. She's seen ospreys and wood ducks here, along with a "steady progression of neighbors and new neighbors, coming in to take advantage of the bounties associated with repeated occupation by beavers."

"Obviously the beavers don't stay permanently," Sue explained, "because they literally eat themselves out of house and home and have to move on until the right trees grow back. But mink, otter, moose, and bear are all attracted to the wetland and marshy habitats, and they're all still here." Sure enough, Sue found a place where a moose had bedded down, leaving a moose-sized oval of twisted and matted vegetation. And she told me about other times when she's seen moose in person, using this very pond to feed and cool off (in fact this is where Sue took the photo of the moose that opens this story). Around a bend, Sue found an otter scent-marking mound, pointing to the fish scales and crayfish shells that the scat contained for proof.

Sue—who spends as much time as she can studying and observing the wild animals of Vermont—has a lot of incredible stories to tell. "A few years ago, on the edge of a nearby river, I watched a mother otter repeatedly

A heron nest in the distance. How many are there? Eight herons in three nests!

A beaver lodge under the heron rookery.

Two river otters.

Forever Wild ~ 159

going out into the water, followed by her three youngsters," Sue related. "She would herd fish into the rocky shallows, and then catch them. And those kids were saying, 'Oh, is that how it's done?' It was really cool!"

I asked Sue what she wanted other people to understand about this place. "Well," she answered, "it's one of those places that embodies an opportunity. It's an opportunity to let nature be the infrastructure here, let nature manage herself here, let nature enjoy the quietude of the remoteness of this place. And that doesn't say someone couldn't come here, much as we're doing today, but for this to suddenly be developed with loop trails would ruin it. Those guys [motioning back up to the herons] wouldn't be happy with it."

"But saying everything I've said about recreation, it might seem to some people that I'm opposed to it categorically. And I'm not. I'm a member of the Green Mountain Club. I think the Long Trail is great. I think the Appalachian Trail is fabulous, and I've hiked much of it. I'm totally OK with all that, and proud of it, in fact. I guess what I hope is that more people can do as we're doing in the Green Mountain Club, for example, and give back. We somehow have to make it better for the animals. And part of making it better for animals is to recognize that their refugia, their quieter habitats, their bigger, wilder places like this—these places really work for them. And it's not appropriate for us to be everywhere."

"So, part of giving back is to say, 'Here's where I won't be. Here's where we shouldn't insist on a lot of recreational infrastructure. Here's where beaver can go ahead and flood the place, or bobcats can be predators and not be judged and condemned by us.' Some habitats should be permanently left wild enough to accommodate and support even apex predators, like wolves and pumas when they return."

We made our way back along an old logging road that the beavers had taken over and flooded. "It does my heart good to see that," Sue let out, unprompted. On our way back we passed wonder after little wonder, like a row of tiny mushrooms popping out of an old chainsaw mark, and ants raising their young inside a tiny pine cone.

We passed a beech tree with old bear claw scars running up its trunk, and I asked Sue one final question. "If the land could talk, what do you think it would say?" Sue looked thoughtful, and then replied, "I guess I just don't really think about it that way. It's probably a shortcoming on my part . . ."

I thought that was the end of her answer, but after a pause, Sue continued, ". . . but, if we could *listen*! The land *is* talking, in its many multitudinous ways, but if we could listen better we'd probably do a better job at living life. We'd learn a lot more about patience, and sharing, and we'd learn to recognize greed for what it is, and waste for what it is. If you think about it, there's no waste in nature. Why is it that there's so much waste in what we do? What does that tell us about ourselves?"

Thinking back now on what Sue said, it reminds me of what I've heard people say about relationships: that an integral component of any relationship is leaving enough space for others to be themselves on their own terms. If you don't, you hurt them. But you also lose the ability to learn from them, and to see the world through a different lens. If we don't leave enough space for nature to thrive on its own terms—to have quiet refuge across a diversity of connected habitats—we'll be both hurting nature and losing an opportunity to learn how to be better people.

Rewilding Firefly Hill

Biodiversity Conservation in a Backyard

ERIC HAGEN

When Charlie Hohn and his family moved into their home outside of Montpelier six years ago, his mower kept getting stuck in the mud. It turned out that a little spring on his neighbor's property was flowing into his yard. As a wetland ecologist, Charlie knew that even small seeps are important for biodiversity and water quality, so he decided to stop mowing. "When we bought the land, the spring was this little thing flowing into the ditch on my neighbor's property. Pretty soon I started planting some native plants down here on my side because it was kind of wet."

Later, one winter when the ground heaved with all of the freezing and thawing, the water found an animal burrow and shifted onto Charlie's land. "I know it's not like the water is literally saying 'Oh, I'm going to go visit that thing that Charlie built,' but it almost feels like the wetland is sucking in the hydrology, and sucking in the components that belong there."

Charlie's soggy lawn has turned into a functioning ecosystem that now helps filter and retain stormwater, and provides a home to new plants and animals. "I'll get rid of invasive plants if I can, and then in the wetland I'll plant native plants if I can get them," said Charlie, explaining how he takes care of his yard. Other native plants have moved in on their own, followed by insects and amphibians. "I'm just trying to heal something in a very small sense with this land—positively influence at least something small."

Speaking of his three-year-old daughter Holly, Charlie said, "My other passion now is that I get to show Holly all of this stuff. I grew up in a concrete wasteland where we didn't have nature and wetlands like this." Turning to the open water, Charlie said, "That's our fishing hole—for pretend, obviously, there's no real fish in there—but Holly will pretend to fish in there, and play in the mud. She invented it herself!"

Charlie said he never wants to move again; he thinks it's nice to be tied to the land. He cherishes sharing it with his daughter and seeing what captures her interest, and following her with that.

"Having her grow up with a wetland next to her house," Charlie said, "how cool is that? I mean maybe she'll be like, 'Ugh, Dad, you and your weird wetland,' but maybe she'll be like, 'Yeah we have a wetland, and it's not some stinky thing, it's really cool!'" Watching the land change under his care and watching his daughter grow up on it are important to Charlie, but he thinks they might be even more important someday for his daughter, whose foundational memories will be made here.

Charlie thinks a lot of our modern environmental problems come from people not seeing themselves as a part of natural systems. For his part, he's taken the same responsive approach with the rest of the yard as he did with the wetland, giving a home to an abundance of diverse life. "I'll wander around in there," Charlie said, talking about how he manages his yard, "especially in the fall or in the spring when stuff's laid down, and I'll yank anything I don't want: aggressive invasive plants like the buckthorn or *Rosa multiflora*." Charlie's approach is like a partnership. "It's not totally hands-off," Charlie said, "but the land mostly just does what it does. I'm just trying to help make it better."

This was all lawn eight years ago, this whole thing. When I was first mowing it I was like "Ugh, it's a pain to mow this, it's too wet." And then I was like, "Well, why am I mowing in a wetland? This is absurd!"

CHARLIE HOHN

Grass skipper butterflies on vetch flowers.

In 2020 the U.S. Fish and Wildlife Service determined that declining monarch butterfly populations warrant endangered species status, but deferred their listing due to a lack of resources.

Taking care of his land has also turned into a sort of game for Charlie, and he's trying to see how many species he can get to live there. Using the online application iNaturalist, Charlie has documented over four hundred species of plants, fungi, birds, mammals, and insects—all in his front and back yard.

Charlie cuts back just part of his yard each year to keep it open, but he carefully times it late each fall so he doesn't harm any monarch caterpillars, and so the milkweed seeds have time to mature and spread.

Charlie explained further. "I'm trying to bring it back such that it looks like a functioning, biodiverse ecosystem, so that it can be positive for biodiversity, for my kids, and for other community members." It might not seem like converting a lawn could make a big difference, but the United States currently has about 40.5 million acres of lawn (which is an area the size of New England), and studies show that lawns are much less biologically diverse than natural ecosystems.[1]

Charlie continued, "Anyone who owns even a little bit of land can change what's there, and can support a lot more different things. I'm also really fascinated by it. I get bored with something that doesn't change, and now this land is something to watch. Just like animals come and go, plants come and go, and the water changes. That's the story behind this, and to me that's way more interesting than having a lawn."

Walking through his yard, the milkweed blossoms looked like a little living fireworks show, and the bees love them, too. "Life supports life, right?" said Charlie, "Like life on Earth supports human life, and it also supports the animals and plants we care about. Sugar maples, or cows, or anything else, they're all supported by other living things. And so it's important to have biological systems intact so we don't all die, or watch our kids die. But also, having robust biological systems intact gives us more options, more possibilities to adapt to a changing world moving forward."

Charlie was explaining how biodiversity helps us in a material sense, but he also thinks there's more to nature than just what it produces for us. "I think there is a spiritual sense to our relationship with nature, too—it's just hard to describe in our culture. It's like being able to feel the ecosystem, and what it's doing. But yeah, I wouldn't even know how to describe that . . . maybe it's like you can tell when you walk into a place what's going on with it to some extent, feel some sort of connection. Or at least I feel like I can, even if I'm making it up."

Charlie has a few other projects on his land. He's built a rain garden to catch the runoff from his roof and driveway, and modeled it after the dams of beavers that cascade down a waterway. He's built a series of berms that he's covered in water-loving native plants. During a rainstorm the depressions behind the berms fill with water. If the first pool fills up, the water flows downhill and gets trapped behind a second berm, then a third, then a fourth.

Alluding to the way that forestlands provide timber, Charlie said that his wetland is also a working landscape, and so is his rain garden, slowing runoff to purify water, provide habitat, and reduce the severity of flooding downstream during large storms. Every little wetland is like a sponge, slowing down the flow of water and spreading out the discharge over a longer time period, making any potential floods less catastrophic. This flood protection role of wetlands was extremely evident during tropical storm Irene, when intact wetlands and floodplains helped save the town of Middlebury, Vermont around $1.8 million dollars by reducing flooding by 90 percent.[2]

His dad named the land around Charlie's home Firefly Hill, and if you visit on an early summer evening you'll see tons of them. But the fireflies are missing in the other yards nearby that are mown. "These species, these

fireflies that are here, aren't just something cool for our kid to see, they're also an amazing story. I think that every organism is a story, an unbroken line back to however and whenever life started."

Talking about a maple tree growing in his yard, Charlie continued, "That maple tells you something really interesting, that for the last 80 years a maple could survive there. Because if for one minute it would have been enough to kill a maple, then it would have been dead. So every plant that has been here for a while has this really interesting story, especially when you gather it all together."

All the plants and animals around us have their own story—stories interwoven with each other, stretching back into the eons—and now we humans are a part of all these stories like never before. May we be a part of them in a positive way.

REFERENCES

1. Chollet, S., Brabant, C., Tessier, S., & Jung, V. (2018). From urban lawns to urban meadows: Reduction of mowing frequency increases plant taxonomic, functional and phylogenetic diversity. *Landscape and Urban Planning*, 180.

2. Watson, K. B., Ricketts, T., Galford, G., Polasky, S., & O'Niel-Dunne, J. (2016). Quantifying flood mitigation services: The economic value of Otter Creek wetlands and floodplains to Middlebury, VT. *Ecological Economics*, 130.

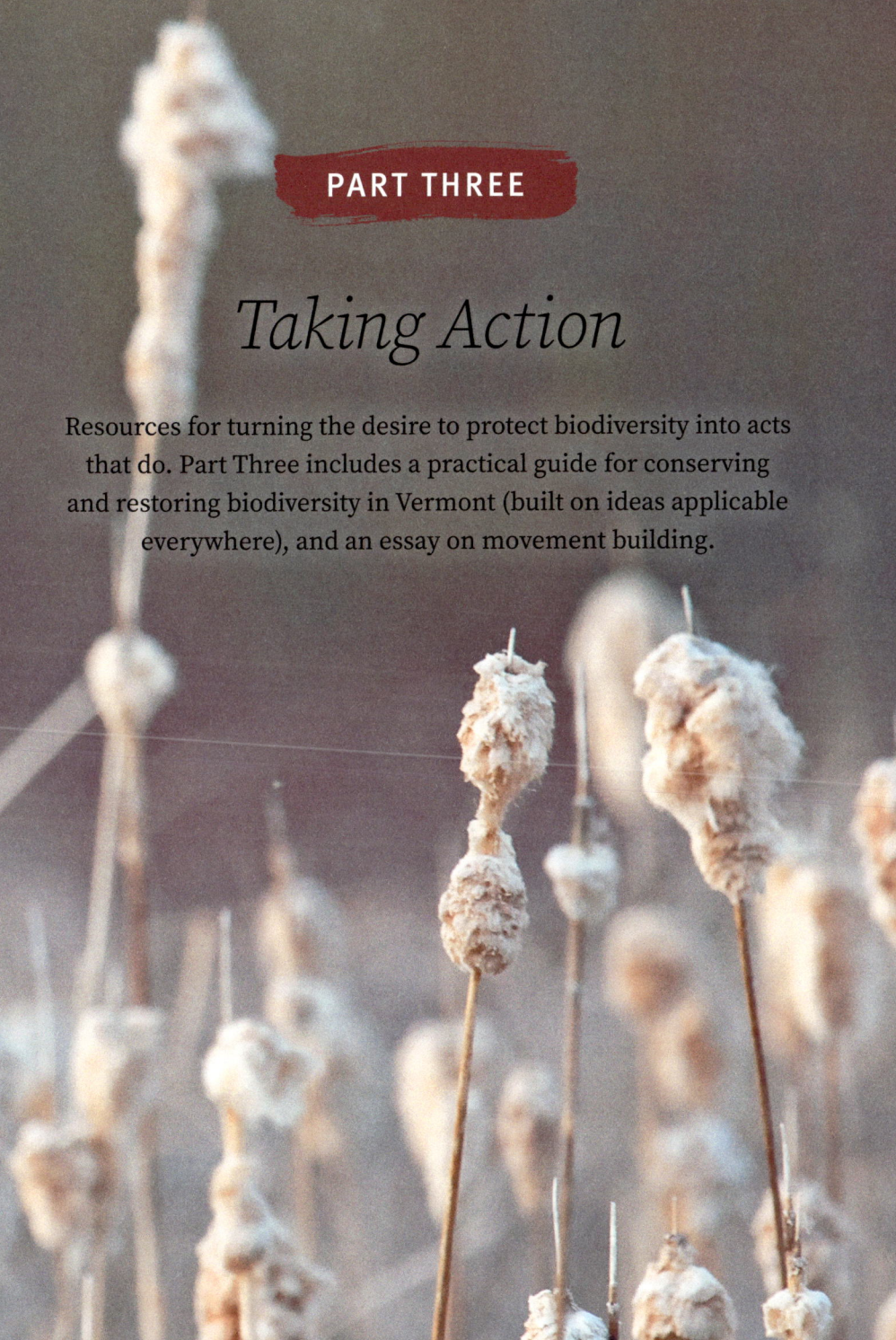

PART THREE

Taking Action

Resources for turning the desire to protect biodiversity into acts that do. Part Three includes a practical guide for conserving and restoring biodiversity in Vermont (built on ideas applicable everywhere), and an essay on movement building.

I wonder how we could begin to recognise . . . that messy stretches of species-rich vegetation with all their attendant invertebrate life are better, just better, than the eerie, impoverished silence of modern planting schemes and fields. I wonder how we might learn to align our aesthetic and moral landscapes to fit that intuition. I wonder.

HELEN MACDONALD

A Vermonter's Guide for Protecting Biodiversity

At Home, in Your Town, and Across the State

ERIC HAGEN

How to use this guide

This guide is designed to give Vermonters clear instructions for directly restoring, enhancing, and protecting biodiversity in the areas that we have the most influence over: our homes, our towns, and our state. For readers outside of Vermont, consider this guide a starting example or a source of inspiration for understanding how to protect biodiversity in your own part of the world.

These straightforward instructions provide a foundation for understanding what you can do to steward biodiversity in different contexts. This guide is organized at two different scales, half-yard and half-town, because we think that these are two scales of conservation where ordinary citizens can make a big difference for biodiversity protection. The information in this guide is also organized by these two scales because different conservation concepts are most applicable when considering differently sized areas. The truth is, though, to best steward biodiversity in the smallest areas, like your backyard, you'll have to understand your context within the wider world. Inversely, in stewarding the whole world we'll have to apply our efforts in the smallest of places. You'll get the most out of this guide if you read the whole thing, and then apply the concepts that seem the most useful for the area that you want to work in.

We use the word "half," as in "half-yard," not in a literal sense, but to indicate that wherever we live and have influence over land use decisions, we need to consider biodiversity and ecosystem function. "Half for nature, half for us" is a good motto, but reality will be more nuanced. Some areas should be wilderness, others highly engineered for people, and many places in between, with the accounting landing somewhere above or below half in different contexts. We would like to argue, however, that even in highly engineered places, like cities and parks, native plantings should be used in order to integrate human spaces into the wider food web, and development should be sited in order to cause the least disruption to animal and plant movement.

Though the information here is presented in brief, there are abundant online and print resources that cover each topic, published by various organizations and referenced throughout this guide. You can access these resources by going to vermontallianceforhalfearth.org/strategies to access an online version of this guide which contains clickable links to all of the resources mentioned.

What is biodiversity?

Biological diversity, or biodiversity for short, is essentially what many people think of as nature. In more technical terms, biodiversity is the complexity of all life on Earth, at all levels of organization. In its simplest form, biodiversity is a measure of the variety of species and their relative abundance. A fuller understanding of biodiversity also includes the genetic diversity within species, interactions between species, and the organization of species across the landscape. This all means that a habitat which supports large, stable populations of more species is more biodiverse. It also means that each habitat and ecosystem uniquely contributes to global biodiversity by being the home of specific species of plants, animals, fungi, and microorganisms.

Right now, the Earth's biodiversity is in crisis. Measured by weight, there is currently only half as much life on Earth as there was before the rise of human civilization.[1] Looking at just our closest relatives, only 4 percent of the mammalian biomass on Earth is wild, while the other 96 percent belongs to either humans or our domesticated species.[2] Likewise, birds have declined by 30 percent across North America in less than 50 years, and 40 percent of the Earth's insect species are in decline and facing extinction.[3,4]

As the abundance of life decreases on Earth, the likelihood of local and global extinction of individual species rises.[5,6] Indeed, a 2019 United Nations report concluded that currently 1 million species on Earth are facing extinction, many within decades.[7] As a whole, humanity is clearly failing to live in harmony with the rest of life, and we need to change that.

Why is biodiversity important?

There are many lenses through which we can see the value and importance of biodiversity.

Instinctively, many of us feel that nature simply has a right to exist and thrive for its own sake, believing that humans don't have to receive any value from nature in order to protect and steward it.

Another perspective is that biological diversity is the foundation of healthy ecosystems, and healthy ecosystems provide valuable goods and services. This seems like common sense. All living organisms depend on other living beings for survival, and we're no exception. Sometimes, though, it's helpful to give common sense a number. According to the World Economic Forum's 2020 Global Risk Report, $44 trillion of the world's economy (over half of global GDP) is moderately or highly dependent on nature and its services. Here in Vermont, healthy ecosystems contribute to our economic well-being by providing clean water, flood protection, pollinator services, carbon sequestration, timber production, and recreational opportunities.

Beyond working to produce our oxygen, capture our carbon dioxide, and clean our water, nature has been shown to contribute to human health in a number of other ways, improving human well-being from cradle to grave. Evidence-based benefits of nature contact include improved birth outcomes; improved child cognitive and motor development; reduced ADHD symptoms; lower risk of psychiatric disorders; better eyesight; decreased stress, anxiety, and depression; reduced diabetes and obesity; improved immune function; reduced risk of death from cardiovascular disease; general improved happiness and life satisfaction; and more.[8]

One last perspective to consider is how we find value in our relationships with nature. Memories are often tied to natural places, along with our sense of identity. If you've ever been away from Vermont for a long time, and then felt your heart swell when you see its forest-covered mountains again—or if you ever have memories spring to life with the changing sounds, smells, and sights of the seasons—then you understand this sort of relational value.

Many of us feel a sense of kinship and friendship with the natural world, or our sense of moral integrity depends on our stewardship and partnership with nature. For many people, nature is a necessary entity for passing on traditions and values (see "A Field that Bloomed," page 129). Protecting biodiversity can be our way of giving back to the ecosystems that support us, while also caring for ourselves, each other, and future generations.

The personal scale of biodiversity conservation

Half-yard, half-school, half-business, half-park . . . even in the smallest of places we can make decisions that help biodiversity. Read Charlie Hohn's story ("Rewilding Firefly Hill," page 163) to see how one Vermonter is rewilding his yard, and use the information below to create your own personal biodiversity hot spot.

Plant natives. Native plants support more of our local insects, birds, and other wildlife. Because of this they are the foundations of food webs and act as biodiversity multipliers. Just like monarch butterflies need milkweed for their caterpillars to survive, many other insects depend on specific

Most of our native caterpillars can eat only specific native plants. This milkweed tussock moth caterpillar is dependent on milkweed for its survival, just like the monarch butterfly caterpillar and 10 other moth and butterfly species. Some plants, like oak trees, support over four hundred species of caterpillars.

native plants to complete their life cycle, and these insects in turn are a food source for other creatures. Native plants also reflect the unique beauty and natural heritage that only Vermont can offer, and planting native plants honors the relationships between plants, animals, and the landscape that have developed over thousands of years.

There are a lot of great resources for finding native plants appropriate to your location and needs:

- Audubon has an online tool, **Plants for Birds**, which helps you find plants native to your zip code. Audubon also has a great set of articles explaining how to make your yard more bird-friendly.

- The National Wildlife Federation has another zip-code specific tool, **Native Plant Finder**, which tells you how many species of butterfly and moth caterpillars each plant supports. This is important because it allows you to increase and diversify (and therefore stabilize) the food source for birds during their critical nesting season.

- The Native Plant Trust has perhaps the most versatile tool, **Garden Plant Finder!**, which lets you narrow your search results to specific plant and habitat attributes, like flower color, size, moisture preferences, and more.

This lesser purple fringed orchid is growing in a sunny Rich Fen natural community, which itself is embedded within a Northern White Cedar Swamp natural community.

If you are trying to restore parts of your yard to a more wild setting, a better approach is to figure out which natural communities would occur on your property if it were in its most natural state. A natural community is essentially a group of plant and animal species that reflect the climatic, geologic, and hydrologic characteristics of the place they are growing (Rich Northern Hardwood Forest and Northern White Cedar Swamp are two examples of the over 90 different natural communities in Vermont).

Wetland, Woodland, Wildland is *the* guide to natural communities in Vermont, and it can help you find which species of plants are most likely to thrive and reproduce the best with the least maintenance in your specific location.[9] In a nutshell, to use the natural communities approach to plant selection you would look at nearby intact natural areas that have similar characteristics to the area you want to restore, and then plant the same plants, along with others that would likely do well there. By restoring natural communities we can provide habitat for specialized components of our biodiversity, and reweave Vermont's natural landscape.

Sourcing native plants. The fastest way to get native plants is to go to a nursery that supplies them, and the Audubon tool mentioned above can help you find local vendors. When possible, it's best to avoid cultivars of native plants because these have reduced genetic diversity.

In nature, plant populations are locally adapted, but also genetically diverse. When rewilding our yards we should try to replicate this by sourcing plants from local seed whenever possible. Some nurseries specialize in using local seed sources. Another option to supplement what you can find from native plant vendors is to start your own plants from seeds that you harvest yourself. A quick internet search provides all the information you need to successfully germinate and tend a huge variety of plants.

Remove nonnative invasives. Nonnative invasive plant species can outcompete our native plants, and therefore decrease local biodiversity where they take over. A website called Vermont Invasives has pictures and management techniques for all of our worrisome invasives. If we remove invasive species we can slow their dispersal to new places. In general, it is best to remove mature plants first so that new seed production is stopped as soon as possible. It's also a good strategy to cut back and manage unwanted plants earlier in the growing season after they send up shoots and deplete the energy stored in their roots, but before they can photosynthesize much and store more. You may have to cut their shoots back repeatedly, pull them, or try other techniques. One effective herbicide-free strategy for "resetting" an area dominated by persistent invasive species is to cut back the plants and then cover the ground with black silage tarp for at least a full growing season. (Silage tarp is made from nonwoven, UV-resistant plastic that doesn't break down into fragments like typical outdoor tarps.) Called "occultation," this technique creates a moist and warm environment that encourages seeds to germinate and the underground parts of plants to send up shoots, but completely blocks sunlight so that these striving plants deplete their energy stores and die. Exercise caution when using this technique to avoid unnecessarily killing native plants, and be prepared to reseed and/or replant the site with native plants (if nearby seed sources are lacking) after the treatment so that the area doesn't become recolonized by invasive species.

Reduce your lawn. Where do you and your family walk and play? A lawn is a biologically impoverished habitat, and should be treated like a specialized tool, not a general default. Shrinking your lawn also reduces the resources and labor that go into maintaining it. Meanwhile, minimize your fertilizer use, and never use fertilizer with phosphorus in it (which leads to algae blooms).

One way to rewild your lawn is to put in new plants. You can also just let it grow, tending the plants that show up by keeping the native plants and removing the nonnatives (iNaturalist is a great tool to help you identify unknown species). You'll probably find some hybrid model between letting nature do its thing and careful landscaping that works best for you.

Messy is beautiful. What we might consider messy fruit and brush, wild critters consider food and shelter. Life is beautiful; sterile landscaping and asphalt, well . . .

Stop using pesticides. Insects play a key role in the wider food web, and pesticides almost always kill more than their intended target. But if you have a mosquito problem, try soaking some hay or yard clippings in a bucket of water and adding a BTI Mosquito Dunk to attract female mosquitoes. This will target only mosquitoes and leave other insects alone.

Leave the leaves. Many moth species, like the declining luna moth, pupate in the leaf litter, and so rely on these dead leaves to survive the winter. In fact, leaf litter provides food and shelter for all kinds of organisms, like fungi, arthropods, and amphibians. Larger animals like birds rely on these organisms to feed themselves and their young.

Dead wood supports new life. Dead wood is like a battery that slowly fuels other forms of life around it. Fungi and beetles eat the cellulose and lignin in dead wood, and in turn these organisms are eaten by woodpeckers and other creatures. Large standing dead trees are also a necessary landscape component for cavity-nesting birds like woodpeckers, chickadees, and owls, and provide denning locations for raccoons, fishers, porcupines, and countless other species. For these ecosystem functions, the bigger the diameter of tree the better.

Wood is like a battery in another way, because it is essentially carbon dioxide stored in solid form. The more carbon dioxide that is locked up in big trees, dead or alive, the less carbon dioxide there is in our atmosphere. It's a win-win for climate and biodiversity.

Structural diversity. Structural diversity means having different three-dimensional shapes on the landscape. You can think of it as having different habitat layers, and habitat diversity is what supports biodiversity. To increase the structural diversity in your yard, make sure to have plants of different heights, like big old trees, young trees or shrubs, and herbaceous plants. If a tree falls in your woods, take what you need for firewood, but otherwise leave it be! The tip-up mounds of fallen trees provide nesting sites for winter wrens, the tangled branches shelter songbirds and young plants, and the rotting trunks make a perfect environment for salamanders. If you want to immediately increase the structural diversity in your yard, consider making a brush pile or a rock pile, or dragging in some interesting logs and sticks. If you're interested, read "Guests at Nature's Table," page 117, to learn how

one Vermonter is increasing structural diversity throughout his gardens.

Mimic old growth. A fun way to think about creating structural diversity is trying to mimic characteristics of old-growth forests. Old-growth forests in our part of the world have big old trees and a well-developed understory, lots of standing and downed dead wood, tip-up mounds, and patches of young forest of different sizes and ages where wind storms created gaps. Old forests also have abandoned beaver ponds that turn into meadows full of flowers. All together, a diverse landscape like this supports a lot of biodiversity. Read "An Unusual Forest," page 143, to see this concept brought to life.

Habitat connectivity. Larger networks of connected habitats support more species. This is because it's easier for animals and plants to disperse through the landscape to find mates and sufficient resources. Whether in town or in the countryside, try to manage your yard so the wildlife habitat within your property is connected to wildlife habitat on neighboring properties and across the road.

Wetlands and riparian buffers. Intact wetlands and riparian areas clean water, reduce flooding, and uniquely support biodiversity (see "Forever Wild," page 153). Naturally vegetated riparian buffers are also ideal corridors for wildlife movement because rivers, streams, and lakeshores stretch long distances across the landscape. Vermont ecologists envision a future where riparian buffers line every stream and lake, and wildlife can move unimpeded across the landscape. View your soggy lawn as an opportunity to revegetate with wetland species, and maintain or restore a buffer of native vegetation along streams (50-foot buffer), rivers, and lakes (300-foot buffers both).

Messy is beautiful, but . . . Some of us don't want our yard to look like it belongs to a haunted house. The good thing is that humanity and the rest of nature don't have to be at odds. With a little creativity and knowledge, we can make yards that are both rich in biodiversity and aesthetically pleasing, which will also help stewardship catch on with others. Here are some tips:

- Carefully mown paths and a sign that says "Bird Habitat" or "Half for Nature" go a long way. Use mown paths to connect areas of interest and use within your yard.
- You can still create a manicured look while using native plants. Use all the conventional landscaping techniques, like edging, designed plantings, and mulching.

- Use the Garden Plant Finder! tool to find native ferns, flowers, shrubs, and trees of different colors and sizes.
- There are more than enough native flowers that are beautiful, and planting flowers that bloom at different times makes for a flower bed or pocket prairie that's easy on the eyes and good for the bees.
- Choose native ferns, spring ephemeral flowers, and other shade loving low plants like American wintergreen as ground covers instead of nonnatives like hostas.
- Rake the leaves off of your lawn, but spread them under nearby taller plants and into the wilder parts of your yard.
- Don't deadhead flowers. Their seeds are important for winter birds, and their hollow stems are important for overwintering native bees. Try bunching them with twine, or, at the very least, cutting them eight inches off the ground.
- You can keep some nonnative plants. It'd be hard to completely give up lilacs and apple trees, and luckily we don't have to. A recent study shows that breeding birds need a landscape with at least 70 percent native plants in order to find enough caterpillars to feed their young.[10] So keep your nonnative ornamentals and food-producing plants to less than 30 percent of your vegetation. Just make sure that they aren't invasive species!

Scaling it up: half-town, half-state

We have the knowledge and the tools to protect biodiversity across the state. What we need is people to participate in municipal planning in order to create robust policy. We should apply the concepts from the "half-yard" section above to the landscaping, maintenance, and revegetation decisions made on town and state property. But at these larger scales there are also new strategies for protecting biodiversity that are fundamentally different. In order to protect biodiversity at larger scales we need to make sure that we protect a sufficient quantity of the full diversity of habitats found in Vermont, that we protect large areas of core habitat, and that all of this habitat is connected so that species can move as the climate changes. Fortunately we have the knowledge and the tools to do this. What needs to happen now is participation in municipal planning and community education so that the science can be translated into policy. Through planning we can meet the uncertain future with town centers that grow in density and vibrancy, and natural areas that stay healthy and connected.

Vermont Conservation Design (VCD). The science at the heart of large-scale biodiversity conservation in Vermont is called **Vermont Conservation Design**. Recently created by the Vermont Agency of Natural Resources and

local partners, VCD identifies the areas in Vermont critical for maintaining our biodiversity and ecosystem function, and also gives guidelines for protecting the unique ecosystem functions in each mapped location. This means that we know which parts of Vermont are the most important for protecting biodiversity, and we also know which sort of human activities are compatible with biodiversity and ecosystem function in those places, and which are not.

The scientists who created VCD defined six *Landscape Components,* six *Community and Species Components,* and four *Vermont Conservation Design Targets* that are uniquely important for biodiversity and ecosystem function. If protected and cared for correctly, these components together would create a network of interconnected lands and waters, effectively maintaining biodiversity and an ecologically functional landscape across Vermont. These components are:

- **Landscape Components**

    Interior Forest Blocks
    Connectivity Blocks
    Surface Water and Riparian Areas
    Riparian Wildlife Connectivity
    Physical Landscape Diversity
    Physical Landscape Blocks

- **Community and Species Components**

    Natural Communities
    Aquatic Habitats
    Wetlands
    Vernal Pools
    Terrestrial Wildlife Crossings
    Riparian Wildlife Crossings
    Rare & Uncommon Species

- **Vermont Conservation Design Targets**

    Young & Old Forest Targets
    Upland Shrub/Forb
    Grassland Refuge Focus Areas
    Grassland Managed Agricultural Land

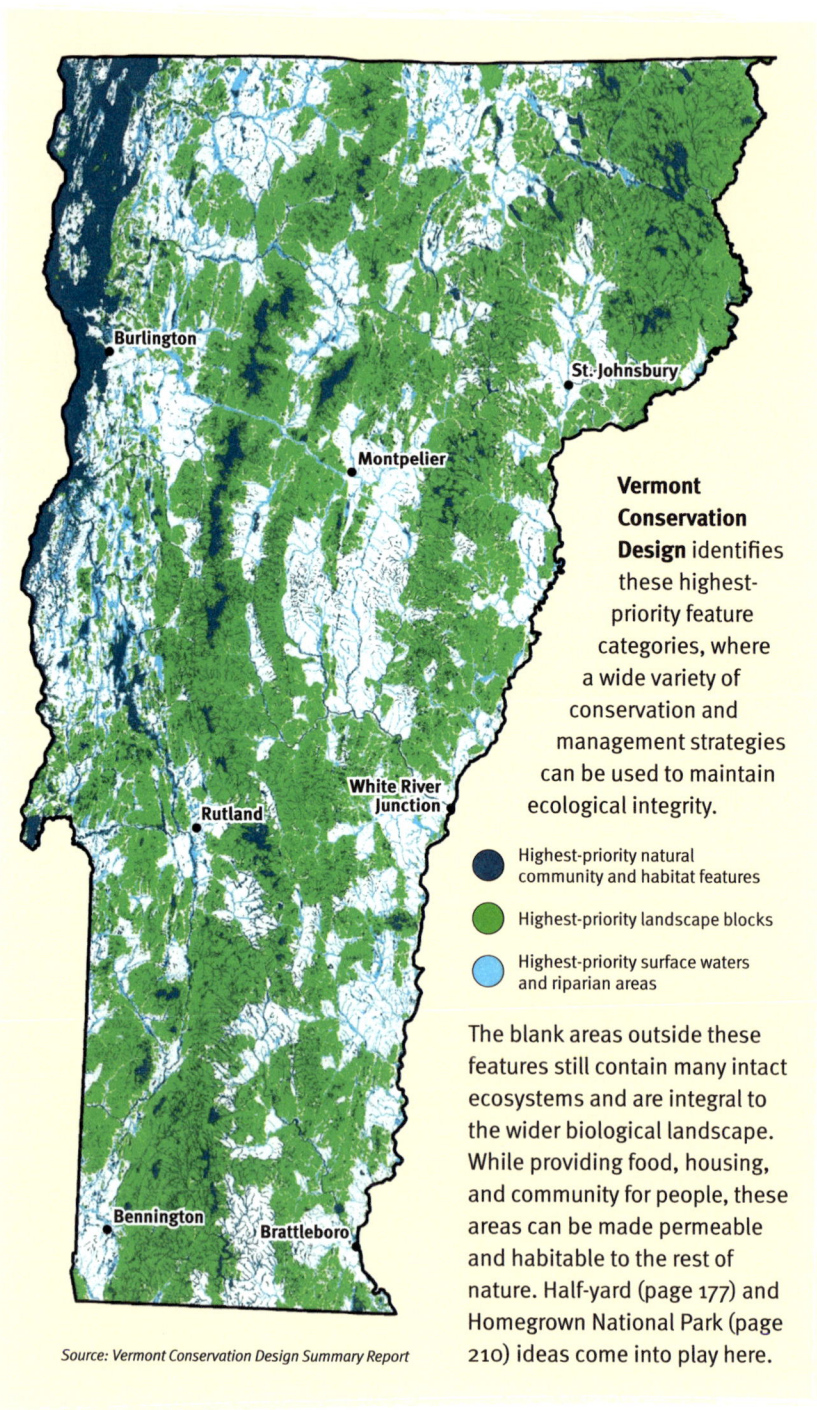

Vermont Conservation Design identifies these highest-priority feature categories, where a wide variety of conservation and management strategies can be used to maintain ecological integrity.

- Highest-priority natural community and habitat features
- Highest-priority landscape blocks
- Highest-priority surface waters and riparian areas

The blank areas outside these features still contain many intact ecosystems and are integral to the wider biological landscape. While providing food, housing, and community for people, these areas can be made permeable and habitable to the rest of nature. Half-yard (page 177) and Homegrown National Park (page 210) ideas come into play here.

Source: Vermont Conservation Design Summary Report

There is an excellent Summary Report of Vermont Conservation Design that explains the science of the plan in greater detail, and reading the "Executive Summary" is a great place to start. Just note that the science of VCD is continuously improving and changing, so the Summary Report uses slightly older language than I describe here. Once you understand the basics, though, it's easy to make the jump.

The authors purposely don't tell us exactly how we should protect the features they identify, but they do offer good suggestions. In this guide we are going to lay the framework for how to follow through with what we perceive to be the most potent and achievable option for groups of ordinary citizens: local planning and policy, which we can use to guide development away from locations that would compromise biodiversity and ecosystem function. The idea is that coordinating development at the town level is more efficient than conserving land parcel by parcel. By making smart development decisions in each region of Vermont we will be able to protect biodiversity across the whole state.

Biofinder. Biofinder is an interactive web map that displays the information of Vermont Conservation Design. The learning curve is a little steep, but the rewards are great. Biofinder allows you to zoom in on any part of Vermont—like your town or land—and see the *Landscape Components* and *Vermont Conservation Design Targets* identified by VCD that are critical for maintaining biodiversity. You can then click on the components and read further descriptions and guidelines for maintaining ecological function. This is critical for conservation and planning commissions, but also useful for guiding management decisions on individual properties and understanding how any property is embedded in the wider ecosystems of Vermont.

Act 171. Vermont towns have "the authority to plan," allowing municipalities to coordinate the management of many of our natural resources. This planning includes implementing nonregulatory approaches like encouraging voluntary best-management practices, and regulatory approaches like creating zoning laws that prohibit development in critical habitats. The recently passed Act 171 provides a crucial opportunity to protect biodiversity by providing a legal and organizational framework to put the science of Vermont Conservation Design into action.

Act 171 is specifically designed to limit habitat fragmentation. It requires regional and municipal planners to identify important "forest blocks" and "habitat connectors," and then to plan development to minimize harm to these areas. This means articulating language about these topics in the town plan, creating a map, and ideally putting policies, recommendations, and actions into place for the plan's implementation. Our job is to become involved in our towns and regional planning commissions to ensure that the identification and mapping process is thorough, and to advocate for effective policies and actions that maintain the functions and values of the identified priority areas. And while Act 171 only addresses habitat fragmentation, the mandate it gives to identify and protect important forest blocks and habitat connectors creates an opportunity to identify and protect other features important for biodiversity.

Get involved. Contact your local planning commission or conservation commission and ask how you can be involved. This could mean talking with members about their plans, encouraging thorough mapping and strong protections, organizing citizen groups to show up to public meetings, or joining a commission when a new opening arises.

In many towns, the Selectboard decides which tasks each commission is responsible for, and they then codify spending and policy decisions recommended to them from these commissions. One powerful strategy to incorporate biodiversity and ecological concerns into town planning is to ask the Selectboard to recommend that zoning regulations are written by a joint committee that includes representatives from both the Planning Commission and Conservation Commission. Another powerful strategy is to run for the Selectboard!

Community context. For someone who cares about biodiversity, it can be very tempting to see zoning and regulation as the silver bullet. Indeed, in communities open to this sort of policy, zoning and regulations could be the most impactful tool for protecting biodiversity available to us. It's worth stressing, however, that these strategies won't be equally accepted in every community, and pushing too forcefully may even backfire. In this guide we are promoting what could be considered a strong path for protecting biodiversity in communities comfortable with regulation, but implementing Vermont Conservation Design is going to require a mixture of regulatory and nonregulatory approaches. Please make your voice heard, but also be a good neighbor, listen to people's concerns, be creative, and organize. The Community Wildlife Program mentioned below is an especially good resource for understanding your community's values and finding the right mixture of regulatory and nonregulatory approaches that work in your context.

Get help. Protecting Vermont's biodiversity is a big commitment, but the reality is that private organizations don't have the money to protect all of the necessary land, and the State wants its laws to reflect local community values. These are both actually good things, but it means that ordinary citizens need to play a significant role in stewarding our own natural heritage. Luckily, we have help.

Various divisions within Vermont's Agency of Natural Resources have created thorough guides to help municipal and regional planners to map and conserve our natural heritage. There is even a program called the Community Wildlife Program whose express goal is to help communities understand conservation science and how to implement it, while reflecting local values. If you ever feel lost, they are a great resource.

Whether you are on a commission or not, it's a good idea to become familiar with the mapping and regulatory tools available to towns so that you can be a more effective advocate for protecting biodiversity. Below is a quick summary of those tools and how towns can use them:

- **Act 171 Guidance**—This document guides planners through a five-step process for "protecting forest and wildlife resources" at the town and regional level.

    Step 1: Assessment of Forest Blocks and Habitat Connectors in Municipal and Regional Plans

    Step 2: Identifying Community Values and Engaging the Public

    Step 3: Identify Community Goals and Policies

    Step 4: Creating Your Future Land Use Map and Policies

    Step 5: Implementation and Action

- **Mapping Vermont's Natural Heritage**—This document helps towns more deeply understand the different landscape components that support our biodiversity. It can be used to inform and complement the assessment of forest blocks and habitat connectors that are the focus of Act 171. Not all of the features necessary to protect biodiversity will be part of forest blocks and habitat features, and this guide can help assess and protect those features missed by Act 171.

    Pre-made maps—There are seven natural heritage maps already created for each town. *Mapping Vermont's Natural Heritage* walks you through each of these maps, explaining what they show and how to use them to protect biodiversity at the town level.

Quick tips. Overlay districts are a type of zoning that trigger heightened review on development proposed within their boundaries. Identified forest blocks and habitat connectors can be used to create overlay districts. If we can get zoning regulations that protect these areas from fragmentation we will go a long way towards protecting biodiversity in Vermont.

When writing zoning policy, try to use strong language like "must" or "shall," rather than weak language like "should" or "may." An example from Act 171 Guidance: "Development that takes place within identified forest blocks shall be located at the edges of the blocks in order to reduce fragmentation of the block by roads, clearing, and development. If there is no land that is physically suitable for development at the edge of the blocks, the development must be located in order to minimize fragmentation of the block."

Example of town plan and Act 171 compliance. In 2019 the town of Wardsboro updated their town plan and town plan maps to reflect Act 171. These documents can be used for guidance while complying with Act 171 in your own community.

Closing considerations

Our efforts to protect and promote biodiversity will be as varied as the landforms, creatures, and people we are trying to care for. To build strong human and nonhuman communities, we'll each have to follow our own curiosities, work together, and find our own conservation niche. This guide was written to provide a solid foundation for these efforts—a framework of effective concepts and strategies for protecting biodiversity at the personal, local, and regional levels.

Take action, enjoy the journey, and thank you.

Go to **vermontallianceforhalfearth.org/strategies** for a version of this guide with clickable links to all of the resources mentioned.

REFERENCES

1. Bar On, Y. M., Phillips, R., & Milo, R. (2018). The biomass distribution on Earth. *Proceedings of the National Academy of Sciences*, 115(25), 6506.

2. Ibid.

3. Rosenberg, K. V., Dokter, A. M., Blancher, P. J., Sauer, J. R., Smith, A. C., Smith, P. A., et al. (2019). Decline of the North American avifauna. *Science*, 366(6461).

4. Sánchez-Bayo, F., & Wyckhuys, K. A. G. (2019). Worldwide decline of the entomofauna: A review of its drivers. *Biological Conservation*, 232.

5. MacArthur, R. H., & Wilson, E. O. (2001). *The theory of island biogeography*. Princeton University Press.

6. Matthies, D., Bräuer, I., Maibom, W., & Tscharntke, T. (2004). Population size and the risk of local extinction: Empirical evidence from rare plants. *Oikos*, 105(3).

7. IPBES. (2019). *Global assessment report on biodiversity and ecosystem services of the Intergovernmental Science-Policy Platform on Biodiversity and Ecosystem Services*. IPBES secretariat.

8. Frumkin, H., Bratman, G. N., Breslow, S. J., Cochran, B., Kahn Jr, P. H., Lawler, J. J., et al. (2017). Nature contact and human health: A research agenda. *Environmental Health Perspectives*, 125(7).

9. Thompson, E. H., Sorenson, E. R., & Zaino, R. J. (2019). *Wetland, woodland, wildland: A guide to natural communities of Vermont*. Chelsea Green Publishing.

10. Narango, D. L., Tallamy, D. W., & Marra, P. P. (2018). Nonnative plants reduce population growth of an insectivorous bird. *PNAS*. 115 (11549–11554). https://doi.org/10.1073/pnas.1809259115

We abuse land because we regard it as a commodity belonging to us. When we see land as a community to which we belong, we may begin to use it with love and respect.

ALDO LEOPOLD

Building Movements to Protect Biodiversity

The Diffusible, the Stewardable, and the Possible

ARVIND SINGHAL AND EVA DUEDAHL

We (Arvind and Eva) met four years ago in Lillehammer, Norway, when Arvind led a seminar at the Inland Norway University of Applied Sciences on the theory of diffusion of innovations, including its potential and pitfalls. As a novice doctoral student interested in transitions to sustainable development, Eva mustered courage to ask: "How can one avoid the folly of diffusing quick-fix solutions?" This became the starting point of a shared, collaborative endeavor to explore how meaningful processes of sustainable change might be facilitated and enabled across research and practice. This collaboration continues as we coauthor this chapter on how to build movements to protect biodiversity. Specifically, we interrogate how the framework of diffusion of innovations, principles of stewardship, and the narrative strategy of entertainment-education can create the enabling conditions for humans and biodiversity to flourish in local, regional, national, and global contexts. We have organized our ideas, narratively, around three headings: (1) the craft of the diffusible, (2) the values of the stewardable, and (3) the art of the possible.

The craft of the diffusible

The craft of the diffusible explores how actionable practices to protect and enhance biodiversity might emerge and spread more widely. Let us illustrate. In 2009, the British newspaper, *The Independent*, launched the Great British Butterfly Hunt (GBBH), inviting citizens to spot 58 species of butterflies, including 56 native and two migrators. The GBBH came in the wake of two unusually wet summers (2007 and 2008) that sent butterfly populations plunging in the United Kingdom. Newspaper readers received a full-color butterfly poster and regular guidance on how to identify, for instance, the scarlet-and-black red admiral or brown hairstreak. Competitions and prizes were announced to honor those who recorded the most sightings of species. GBBH got a wide swath of locals and visitors, families and schools, and neighborhoods and municipalities involved in the butterfly hunt, raising awareness and appreciation of nature, and creating the enabling conditions for humans to interact with nature. GBBH's success notwithstanding, invitational, purposeful, and long-running human-nature interaction initiatives are highly uncommon.

Also rare are science teachers—like Sandra Fary in Vermont, USA—who have designed, have implemented, and are spreading a hands-on, field-based conservation curriculum for elementary, middle, and high school students (see "High Flyers," page 67). Fary's students, or students of dozens of teachers who have been inspired by Fary, do not sit much in air-conditioned classrooms learning about biology and ecology through PowerPoint. Rather, they are investigating life in their backyards, recognizing local birds by sight and sounds, or undertaking a BioBlitz—a biological census of every living organism in a chosen site. Armed with nets, bug boxes, vials, and their iPads, the eager young scientists take pictures of the life they encounter, knowing it will find a place in the iNaturalist.org site, contributing to the biodiversity database for a neighborhood, county, or state.

The pedagogical practices implemented by Fary and other enthusiasts, like Alicia Daniel (see "Saving the Forest by Learning the Trees," page 85), Trish O'Kane, and Sean Beckett (see "High Flyers," page 67), have spread across schools, school districts, universities, and a variety of nature centers and conservation organizations in Vermont. These practices are simple, experiential, and carried out in association with others. A morning activity may involve students finding their short "sit spots"—spots where students sit in nature for five minutes with their antennas tuned to absorb the hyper-local sights, sounds, smells, and textures. An afternoon activity may involve working with others on establishing a pollinator garden to attract birds, bees, bats, and butterflies. Such nature-centered learning practices have led middle and high school students to encourage their families to adopt the half-earth mission—that is, devote more of their land to nature.

Fary is an influential diffuser of a rich biodiversity curriculum that includes dozens of field-based actionable practices. *Diffusion* is the process through which an innovation—a new idea or practice—is communicated through certain channels over time among the members of a social system.[1] Diffusion studies have shown a predictable over-time pattern when an innovation spreads—the familiar S-shaped cumulative adoption curve. The take-off in the S-curve is due to the engagement of early adopters, particularly opinion leaders, in talking about and modeling use of the innovation for others to hear and see and try.[2] As one would expect, BioBlitzes have sparked a multitude of new conversations among diverse groups of Vermont citizens about nature and biodiversity in schools, homes, neighborhoods, communities, organizations, and policy forums. The adoption of new ideas and practices—whether in protection or restoration of biodiversity—occur more rapidly and widely when new conversations are spurred, when new positive experiences are shared, and where the adopters—teachers—or other adopting units—for example, families, schools, school districts, universities, and others—have agency to implement programs and policies.

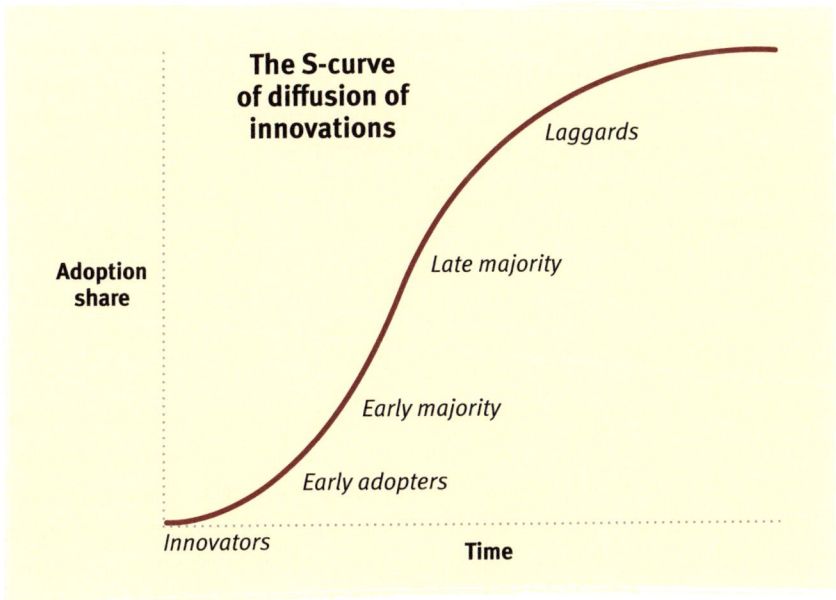

Fary represents both an innovator—willing to take risks, initiate, and experiment—and a highly influential opinion leader, one who holds sway with school and district administrators, and with peer teachers. In turn, these administrators and teachers represent early adopters and influence others in their peer network. Equally influential are the hundreds of receptive students and their families. So goes the diffusion. Thanks to the relationships and conversations Fary has forged with other teachers and conservationists, 10 schools joined in a fall 2019 BioBlitz. Hundreds of students recorded over 2,500 observations on iNaturalist, leading to the identification of 600 different species found on the school campuses. One of the exciting findings was the sighting of *Parancistrocerus leionotus*, a species of potter wasp, by a student at Lamoille Union Middle School. (This is only the second record of this species on iNaturalist in New England.) Even during the COVID-19 pandemic the group expanded the effort in 2020, inviting schools and all Vermonters to join in the Vermont Spring Backyard BioBlitz. Some one thousand residents recorded 10,648 observations. Seeking to further support such actions, one of the BioBlitz sponsors, North Branch Nature Center, collaborated with the teachers to create the Vermont Biodiversity Institute for Educators.

How to diffuse the craft of the diffusible? The case of the Great British Butterfly Hunt and the pioneering work of Fary and her colleagues in Vermont espouse the craft of the diffusible in the design of creative campaigns, curricula, and field-based experiences that invite and engage citizens in diffusing simple and doable biodiversity practices. Notably, the cases illuminate how innovative ideas and practices to protect biodiversity diffuse not by themselves but always in community with others. In this regard, innovations equipped for protecting biodiversity diffuse more quickly and widely when shared through hopeful and generative conversations and engagements with nature.

The values of the stewardable

The values of the stewardable concern questions about how meaningful human-nature relationships can be purposely nurtured. Stewardship means "caring and loyal devotion to an organization, institution, or social group,"[3] where care can be understood as "looking after" something or someone, that is, nature.[4] While stewardship does not dismiss personal motivations, it believes that benefits occur when other interests are put above one's own and when actions are pursued that generate intrinsic rewards.[5] Charlie Hohn (see "Rewilding Firefly Hill," page 163) stopped mowing his waterlogged lawn, got rid of invasive flora in his backyard, and planted native species to enhance his wetland's biodiversity. As a steward, Charlie's motivations are intrinsic—his gratification comes from making a positive impact on his children, community members, and the planet. How might we engage in such stewardship, take stewardship actions, and enable and inspire stewardship with others and nature?

George Schenk, owner of Lareau Farm and American Flatbread shows us what it can mean to be a steward. In 1985, Schenk established the American Flatbread restaurant, featuring food that contributes to the health of both humans *and* nature. Visitors to neighboring Lareau Farm will notice an assortment of bug houses, fence posts, and brush-pile tepees—all

Look what I found! A third-grade student participates in a school BioBlitz in Warren, Vermont.

purposely designed to create natural habitats for microorganisms, birds, and wildlife to thrive. Schenk's farm reflects nature as interconnected where no action is too small or too local. When humans engage and care about nature in seemingly small and hyper-local ways, ripples sometimes turn into cascading regional, national, and global waves. The American Flatbread restaurants, imbued with Schenk's reverence for food and nature, already dot the landscape of several northeastern states and are making inroads in the southwestern United States and Canada.

The story of Brad Norman, founder of ECOCEAN, an Australian NGO dedicated to whale shark research and conservation, further illustrates this point. In 1995, while swimming recreationally with the spotted whale sharks off the Ningaloo Reef in Northwest Australia, Norman began to photograph these large (up to 18 meters in length) ocean fish. The spot patterns on a whale shark serve as their fingerprint for no two are alike. By comparing his pictures across time, Norman was using his naked eye to document the whale shark population and track sharks that were regularly returning to the Ningaloo Reef. As swimming with the whale sharks and photographing their spots behind their pectoral fin caught on with others and notably tourists, Norman involved NASA scientists to enable the photo-identification of whale sharks.[6,7] Each photograph, marked with a spatial-temporal stamp, is uploaded to a global library of images to track the patterns of whale shark migration, their life cycles, feeding and breeding habitats, and biology and ecology.

What began as recreational swimming with whale sharks for one individual in a small coastal patch in Australia has now sparked dozens of sustainable tourism operations including the waters of Isla Holbox in Mexico, Tofo Beach in Mozambique, and Donsol Bay in the Philippines. The diffusion of such caring stewardship efforts is important because whale sharks are an endangered species—their populations have declined more than 50 percent in the past 75 years. The knowledge gained from swimming, clicking, and uploading of photographs "makes a vital contribution toward the conservation of the whale sharks and the sustainability of the tourism businesses that rely on them."[8]

Back on the Lareau Farm, Schenk does more than make flatbread (or as he calls it, "pizza with integrity"), he readily shares with others what he is doing: "One of my goals here is to show the smallest, simplest things that anybody can do. You don't need a degree in biology to make up a pile of rocks, but that pile of rocks is going to create habitat for soil invertebrates" (oce

"Guests at Nature's Table," page 117). Little kiosks on the Lareau Farm explain Schenk's farming practices and what they do to enhance biodiversity. He lets community members use some of his farmland for their little vegetable gardens, guiding and inspiring their understanding of how humans and nature, food and nutrition, and healing and restoration go together.

In the last week of August 2021, the collaborative efforts of the Vermont Alliance for Half-Earth, Vermont Natural Resources Council, and other Vermont conservation organizations made possible a community workshop and dinner at Schenk's farm with renowned conservation scientist Doug Tallamy. One hundred fifty attendees learned about hands-on steps that one can take to provide spaces and healthy habitats for Vermont species. Similarly, Brad Norman's ECOCEAN works closely with tour operators, fisheries, and environmental conservation organizations around the world.

What can thousands of tourists in different parts of the world or dozens of concerned citizens in Vermont do for preserving the biodiversity of our oceans and lands? Together, they inspire a global community of caring stewards. As Norman notes: "As a scientist, I can be in one place, one day of the year. But thousands of tourists . . . serve as our research assistants. Their input is a major part of our global monitoring system."[9] The above cases inspire us to consider how stewardship actions and alliances emerge when the enabling conditions are created for others to collaborate in creating healthy human and natural environments.

Recognizing that biodiversity protection cannot be achieved by any one individual, organization, or business in isolation, the unfolding story of the Vermont Alliance for Half-Earth (see "Half-Earth Vermont," page 101) shows the strengths of collaborating with others—educators, conservationists, legislators, business owners, farmers, and naturalists. This book emerged from a conscious effort to build and diffuse new stewardship alliances. Stewardship movements are nurtured when one brings into play the distinct perspectives, knowledge, and capabilities of a diverse set of stakeholders and citizens.

Bringing together notions of stewardship and collaboration[10,11] urges us to consider the values of nature more fundamentally. Here it is useful to distinguish between three approaches to understanding the value of nature: (1) Instrumental: nature is valued for the benefits it renders to humans, and is simply a means to an end; (2) Intrinsic: nature has value on its own, independent of humans; and (3) Relational: value emerges through relationships with nature where the relationship itself is of value.[12] As explored elsewhere in this book (Foreword, "A Rewilding Story," "Adventures in Forest Carbon") valuations of nature are predominantly guided by a strong belief in human supremacy in which nature is perceived as an instrumental resource for human benefit.

The above examples, however, emphasize that to steward the values of the stewardable requires us to move beyond selfish interest and gain. The practices of George Schenk's Lareau Farm, the Vermont Alliance for Half-Earth, and Brad Norman's ECOCEAN arouse us to appreciate that stewardship actions are not performed in isolation but rather emerge and spread when meaningful and sustained connections are built with nature and people. This distinction is vital.[13] If we do not see ourselves as a part of the wider ecological and planetary system, it is difficult to appreciate why we should care for protecting its biodiversity beyond the benefits it renders to us. Notably, relationships and loving partnerships with nature can be rediscovered, learned, nurtured, and passed on between and across generations (see "A Field that Bloomed," page 129). From this perspective, new or changed relationships can be forged and spread when we ask not what nature will do for us humans, but rather what we can do with others and

Contributed by swimmers at sustainable tourism operations, photos of spot patterns underpin ECOCEAN's whale shark conservation efforts.

with and for nature. Herein also lies a fundamental shift from instrumental values to relational values where stewardship actions and movements are seeded in local contexts but nonetheless are global in outlook.

The art of the possible

The art of the possible asks how a wider public might become interested and engaged in protecting biodiversity. The author Neil Gaiman (paraphrasing G. K. Chesterton) writes, "Fairy tales are more than true: not because they tell us that dragons exist, but because they tell us that dragons can be beaten." This pithy sentence celebrates the narrative potential in stories. Stories, real and fictional, have the power to attract our attention, spark our imagination, and paint new scenarios of hope in transforming our realities. They can demonstrate how "monsters" can be overcome. We contend that when it comes to protecting biodiversity, stories represent the art of the possible.

Tom Butler ("A Rewilding Story," page 17) asks: "What story is inclusive and attractive enough to inspire millions or even billions of people to put themselves into it?" He proposes the story of *rewilding*, arguing that "in a time of loss and diminishment for the diversity of life," rewilding represents the meme for more beauty, abundance, and equity among all living creatures. He contends that rewilding is universally resonant: "almost any of us can put ourselves into that story, either through individual or collective action." He notes the potentiality of rewilding in the restoration of Scotland's Caledonian Forest, creation of prairie preserves for bison in Colorado, and the transformation of the great Iberá marshlands in Argentina from "a highly threatened natural area" into a world-class destination for wildlife-watching and preservation.

A story that is inclusive, attractive, and universally resonant also needs to have aspirational milestones that one can work toward. Rewilding, in this sense, is both a journey and a destination for, as Butler argues, it aligns perfectly with emerging frameworks for global conservation including "30 x 30" (protect at least 30 percent of the planet by 2030) and the more ambitious half-earth goal articulated by E. O. Wilson.

How can the rewilding story go global—that is, reach millions, if not billions? How might a new language, a new vernacular, a new set of memes about biodiversity protection be proposed, provoked, and evoked? How might humanity vanquish the monstrous loss of biodiversity and move toward more wilderness, more abundance, and more equity among all living creatures.

Here the narrative strategy of entertainment-education (EE) holds great potential. EE is a social and behavior change communication strategy that leverages theories from multiple disciplines to inform the creative production and dissemination of narrative-based interventions.[14,15] To understand how the EE strategy can be leveraged, consider the narrative potential of events in April 1973 in Mandal village in Utterakhand, India, when a group of women spontaneously "hugged" trees to prevent them from being felled. In the following years, more than a dozen confrontations between women and lumberjacks occurred, enshrining forever the new meme of "tree hugging" in conservation parlance. Stories and photographs of women's bodies interposed between the trees and the axes of timber cutters spurred interpersonal and written conversations in neighboring communities.[16] The *Chipko* ("hugging") movement spread rapidly as it was led by Chandi Prasad Bhatt and Sunderlal Bahuguna, two charismatic local activists and influencers—both deeply inspired by Mahatma Gandhi. Well-networked with journalists, Bhatt and Bahuguna each wielded a prolific pen and wrote with ease in both Hindi and English, mobilizing the rural citizens and urban elites.

While the Chipko movement unfolded in real time in India, its narrative appeal was no less than epic and mythical. The notion of "cut me down before you cut down a tree" brought a new humanized morality to take on abstract environmental monsters. As stories about Chipko were told and retold, and its legendary and mythical attributes found utterance on the wings of animated conversations, it captured the imagination of diverse audiences. Not surprisingly, the movement was soon co-opted, shaped, and popularized by local and global journalists alike, grassroots activists, environmentalists, Gandhians, spiritual leaders, politicians, social change practitioners, and feminists. *Bhagwad kathas* (large prayer meetings) were organized in forest areas of Uttarakhand, emphasizing that God resides in every living being, and hence to protect the trees was a sacred act. The feminist movement popularized Chipko, pointing out that poor rural women walk long distances to collect fuel and fodder, and thus are the frontline victims of forest destruction. The Gandhians accentuated the Chipko movement through symbolic protests such as prayers, fasting, and *padayatras* (ritual foot-marches). Over time, the "Chipko" meme of taking a personal and collective stand against the destruction of nature climbed rapidly on the public and policy agendas of the regional, national, and global publics.

Notably, after Sunderlal Bahuguna undertook a five-thousand-kilometer trans-Himalayan march in the early-1980s, Prime Minister Indira Gandhi legislated a 15-year ban on felling of green trees growing over one thousand meters above sea level in the Himalayan forests. This decree was then extended to the tree-covered forests in other regions of India.

The stories of George Schenk, Brad Norman, Sandra Fary, and the memes of rewilding, stewardship, and Chipko, lend themselves as ready fodder for melodramatic treatment, that is, the portrayal of engaging storylines with embodied conflict between values of protagonists and antagonists. While the EE strategy has been implemented in the form of conflict-laden melodramatic serials in regional and national contexts to address air pollution, encourage tree planting, and save wildlife habitat, these efforts have been ad hoc and isolated.[17] Imagine if a series of coordinated narrative interventions—real and fictional—were created and disseminated to guide and inspire stewardship actions for biodiversity protection.

The art of the possible is about evoking our ability to collectively reimagine the current state of biodiversity. It is about crafting transformative stories and shared narratives equipped at conquering current "environmental monsters" in a quest for protecting biodiversity with others.

In closing

While there is no telling what actions for biodiversity protection will cascade beyond the examples of this chapter, the stories we have told of small, locally situated actions may radiate across time and space, enabling bigger global movements of stewardship. When elementary, middle, and high school students in Vermont begin to convince their parents to "rewild" a chunk of their manicured lawns, it may inspire and trigger a variety of cascading actions by ordinary citizens.

The craft of the diffusible informs how new ideas and practices can be purposely crafted and diffused rapidly by engaging a wide spectrum of potential adopters, spurring rich and hopeful conversations and experiences, and inspiring individual and collective actions in existing and cascading social networks. The values of the stewardable urge us to reconsider the values of nature and to nurture sustained meaningful relationships with other people and the rest of nature. The art of the possible stirs us to reimagine shared stories of biodiversity, knowing they hold the potential to seed new ideas, possibilities, and actions.

REFERENCES

1. Rogers, E. M. (2003). *Diffusion of innovations*. (5th ed.). Free Press.
2. Dearing, J. W., & Singhal, A. (2020, September). New directions for diffusion of innovations research: Dissemination, implementation, and positive deviance. *Human Behavior & Emerging Technology*.
3. Neubaum, D. O. (2013). Stewardship theory. In E. H. Kessler (Ed.), *Encyclopaedia of management theory*. Sage Publications, p. 769.
4. Enqvist, J. P., West, S., Masterson, V. A., Haider, L. J., Svedin, U., & Tengö, M. (2018). Stewardship as a boundary object for sustainability research: Linking care, knowledge and agency. *Landscape and Urban Planning*, 179(25).
5. Neubaum, Stewardship theory.
6. Duedahl, E., & Singhal, A. (2020). "Picking our oysters" and "swimming with our whales:" How innovative tourism practices may engender sustainable development. *SEARCH Journal of Media and Communication Research*, 12 (1).
7. Hughes, M. (2013). Ecocean: Conservation through technological innovation. In J. Liburd & D. Edwards (Eds.), *Networks for innovation in sustainable tourism: Case studies and cross-case analysis*. Tilde University Press.
8. *Ibid*, p. 26.
9. The Naked Scientists. (2010, August). Whale sharks—stars of the sea: Interview with Brad Norman, Ecocean. *The Naked Scientists*.
10. Liburd, J. (2018). Understanding collaboration and sustainable tourism development. In J. Liburd & D. Edwards (Eds.), *Collaboration for sustainable tourism development* (pp. 8–35). Goodfellow Publishers.
11. Liburd, J., & Becken, S. (2017). Values in nature conservation, tourism and UNESCO World Heritage Site stewardship. *Journal of Sustainable Tourism*, 25(12), 1719–1735. https://doi.org/10.1080/09669582.2017.1293067
12. Chan, K. M., Balvanera, P., Benessaiah, K., Chapman, M., Díaz, S., Gómez-Baggethun, E., et al. (2016). Opinion: Why protect nature? Rethinking values and the environment. *Proceedings of the National Academy of Sciences of the United States of America*, 113(6).
13. Liburd, J., Blichfeldt, B. S., & Duedahl, E. (2021). Transcending the nature/culture dichotomy: Cultivated and cultured world class nature. *Journal of Maritime Studies*, 20(3), 279–291. https://doi.org/10.1007/s40152-021-00229 y
14. Singhal, A., & Rogers, E. M. (1999). Entertainment-education: A communication strategy for social change. Erlbaum.
15. Wang, H., & Singhal, A. (2021). Theorizing entertainment-education: A complementary perspective to the development of entertainment theory. In P. P. Vorderer & C. Klimmt (Eds.), *Oxford Handbook of Entertainment Theory*. Oxford University Press.
16. Singhal, A., & Lubjuhn, S. (2010). Chipko environmental movement media. In J. D. Downing (Ed.), *Encyclopedia of social movement media*. Sage Publications.
17. Reinermann, J., Lubjuhn, S., Singhal, A., & Bouman, M. (2014). Entertainment-education for sustainable llfestyles: Storytelling for the greater, greener good. *International Journal of Sustainable Development*, 17(2).

Can we learn to love the Earth enough to change?

TERRY TEMPEST WILLIAMS

Afterword

DOUG TALLAMY

Recently, Curt Lindberg invited me to Vermont to talk about my vision of how E. O. Wilson's half-earth concept might become a reality. It was an inspiring visit. I was honored to meet authors Prucia Buscell, George Schenk, Liz Thompson, and, of course, Curt himself, and I was able to spend time with other members of the Vermont Alliance for Half-Earth, including many of the leading conservation activists in Vermont who have helped to make their state a leader in conservation policy and actions. Yet despite all of the willpower, smarts, and talent behind the half-earth initiative in Vermont and elsewhere, the elephant remains in the room. How can we preserve half of Planet Earth for nature when we humans have already overrun nearly all of it?

I still remember the first time I held Wilson's book, *Half Earth: Our Planet's Fight for Life*, in my hands. Perhaps the most eminent conservation biologist of our time was telling us we must preserve half of the planet. Music to my ears, but how could this be possible? Currently, nearly 50 percent of terrestrial Earth is in some form of agriculture, and 7.8 billion of us, with all of our roads, airports, buildings, infrastructure, waste, and excessive exploitation of all types, are crammed into the other half. Wilson himself spent little time talking about how to preserve half of our planet. Rather, he focuses on the science that clearly says we must allocate half of the Earth for nature, or we will lose all of nature and with it our life support.

This volume has offered solutions to the half-earth dilemma at all scales; from Tom Butler's grand visions of rewilding areas not yet completely transformed by humans; to Jim McCullough's willingness to preserve nearly all of his land in a natural state; to Charlie Hohn's example of putting away the mower to restore firefly populations in his yard. Solutions, however, depend on definitions. When we save the Earth for nature, what exactly do we mean? In many readers' views, nature is wildness, full of charismatic megafauna, including the apex predators so necessary to self-sustaining ecosystems. This definition fits well with rewilding goals that seek to establish large swaths of land in which humans and their detritus are largely absent. Moreover, true rewilding projects are realistic options for places like Vermont because so much of the state remains relatively untrammeled. But rewilding in its purest form would exclude much or all of the land in many other states, as well as the hands-on participation of people who live in densely packed cities and suburbs: some 82 percent of U.S. citizens.

I propose a different definition of nature that is more inclusive and will enable us to achieve much of Wilson's dream, not only on half of Planet Earth but on most of it. What if we define "saving nature" as "saving ecosystem function"? Ecosystems can function well, that is, be highly productive and stable; poorly, with little productivity and stability; or not at all. If we strive to improve ecosystem function everywhere, we bolster the life support that nature provides us day in and day out in the form of ecosystem services, and we also conserve biodiversity, because it is the plants and animals around us that run the ecosystems on which we all depend. Moreover, by improving the ecological integrity of lands between parks, preserves, and large rewilding projects, we build "biological corridors" that connect areas high in species diversity with each other, something ecologists agree is necessary not only for the continued existence of long distance migrants, but for all types of biodiversity.

In 1955, Robert MacArthur suggested that ecosystem productivity and stability is a function of the number of species within that ecosystem: as the number of plants and animals increases, so does the ability of that system to produce services beneficial to all life.[1] Unfortunately, the converse is also true. As ecosystems lose species, they lose productivity and stability. In essence, MacArthur proposed that life begets more life. Robert MacArthur died tragically young, but in the decades since his death, numerous studies have demonstrated that his predictions were correct.

The upshot of MacArthur's insights is that to create ecosystems healthy enough to sustain our enormous human population *without sacrificing most other species*, we need to protect the few ecosystems that remain rich in species, but we also must return species to the many, many ecosystems we have degraded over the years. And here's the catch: we need to do this everywhere! Not just in parks and preserves, not just in the wilder spaces of the planet, and not just on half of Planet Earth. Everywhere. No more "humans here; nature someplace else." We now must live with, instead of apart from, the natural world everywhere we can. Why? For selfish reasons! We need the life-support services produced by high-functioning ecosystems everywhere so that we humans can persist on the finite planet we call Earth.

So, how do we increase the number of species and thus improve ecosystem function in our yards, on our roadsides, at our airports, on the landscapes surrounding our corporations, schools, houses of worship, and yes, even within various forms of agriculture? This is not as hard as you might think. All we have to do is put the plants back; and not just any old plant, but the native plant species that support the food webs that, in turn, support the animal diversity so vital to ecosystem function. Over the past 15 years I have written extensively about why native plants—those species that share an evolutionary history with the plants and animals around them—support far more animal species (both invertebrates and vertebrates) than do plants from other continents and even from other bioregions within North America. But I and my students at the University of Delaware have also found that native plants themselves differ widely in their ability to support food webs by passing on to animals the energy they harness from the sun. Just 14 percent of our native plant species support 90 percent of the caterpillars in North America, which happen to be the most important component of terrestrial food webs.[2,3]

I call these hyperproductive natives "keystone species"; like the keystone in a Roman arch, they are essential to supporting complex, stable food webs. Keystone plants are the two-by-fours in the ecological houses we must all build in our yards. They are essential to holding those houses up. We cannot build a house out of wallpaper—the ornamental plants chosen for aesthetic value rather than ecological contributions. Plant choice, then, is a bit more nuanced than simply favoring native plants over non-native species. We have greater success when we favor our most productive native plants. The concept of keystone plants streamlines our plant choice options, as long as we know which plants can serve as keystones.

Fortunately this information is available online through the National Wildlife Federation's Garden for Wildlife (www.nwf.org/Garden-for-Wildlife/About/Native-Plants/keystone-plants-by-ecoregion).

We will also need to use more plants than our traditional sparsely planted landscape designs have called for in the past. Lawn has served as both a default landscape and a status symbol for over a century. The result of our fascination with this ecological dead-scape is that the United States now has 40 million acres of lawn—more lawn than the area of all of New England. I have fantasized for years now about reducing the area in lawn by 50 percent. That would give us 20 million acres to create a new national park that I am calling Homegrown National Park, since much of it will be created by cutting our residential lawns in half. At 20 million acres, Homegrown National Park will be the largest park in the country, larger than the combined areas of all of our major national parks.

But to change our current landscape designs as I have described, we will need to change our cultural relationship with nature. Agrarian and extractive economies have typically fostered an adversarial relationship with nature. Beat back the natural world before it destroys our crops, floods our homes, or stands in the way of economic "progress." When most of the planet was rich in biodiversity and most ecosystems functioned well, this "Earth is for people" attitude seemed to work. The dominance and exploitation of nature was destructive locally, but not globally. Now, however, we dominate and therefore destroy the natural world nearly everywhere, with the result that we humans have created the sixth great extinction event in the history of the planet. We can continue to bite the hand that feeds us. We can continue to destroy the life-support systems on which we all depend. We can ourselves join the legion of species being obliterated from the Earth—or we can change our adversarial relationship with nature to a collaborative one. After all, we are products of nature, totally dependent upon nature's largess. We will continue to exist only if we recognize this fact and learn to coexist with the natural world, in the same place, at the same time.

REFERENCES

1. MacArthur, R. (1955). Fluctuations of animal populations and a measure of community stability. *Ecology*, 36(3).

2. Janzen, D. H. (1988). Ecological characterization of a Costa Rican dry forest caterpillar fauna. *Biotropica*, 20(2).

3. Narango, D. L., Tallamy, D. W., & Shropshire, K. J. (2020). Few keystone plant genera support the majority of Lepidoptera species. *Nature Communications*, 11(1).

Acknowledgments

A large core of conservation advocates in Vermont made this book possible and contributed to its contents. We benefited from the skills of many members and advisors of the Vermont Alliance for Half-Earth. Some, like Sandra Fary and George Schenk, were featured in stories and essays. Others lent their talents in expected and unexpected ways. Sean Beckett and Steve Shepard contributed many fine photographs. George Schenk and Steve Shepard penned essays. Alicia Daniel wrote about some special natural-history educational offerings in the state. The history of conservation science in Vermont was the topic of the essay Liz Thompson authored.

Other contributors emerged through relationships developed in the course of the alliance's work. A connection with Northeast Wilderness Trust led us to Tom Butler, who wrote the chapter on rewilding. Tom in turn connected us with Annie Faulkner, who wrote about carbon and forests, and Kevin Cross, who became the designer of the book. Andlea Brett, featured in one of the stories, offered to write a reflective essay. Relationships with two contributors—Joe Roman and Doug Tallamy—came about through the alliance's community educational efforts. Arvind Singhal, who has worked for many years with one of the editors (CL), invited a colleague from Norway, Eva Duedahl, to join him in writing the chapter about social movements. Another long-standing connection led to the involvement of journalist Prucia Buscell.

Our appreciation goes to all of these contributors, who did more than offer up their own work. They joined in a very collaborative process. Together they shaped the tone and content of the volume, served as readers of one another's drafts, and joined in the challenge of selecting a fitting title for the book.

Special thanks go to Charlie Hohn, Sue Morse, Lucy and Jim McCullough, Ethan Tapper, George Schenk, and Andlea Brett for allowing their stories about their relationships with the land to be told. Our special thanks also go to Jennifer Esser and Mary Elder Jacobsen for their keen proofreading.

The Lintilhac Foundation, Vermont Natural Resources Council, Northeast Wilderness Trust, American Flatbread at Lareau Farm and Forest, and several of the book's authors provided financial support to make publication of *Our Better Nature* possible. For this we are grateful.

About the Contributors

Sean Beckett is the Director of Natural History Programs at North Branch Nature Center in Montpelier, Vermont. He is also a wildlife photography guide and teacher who brings travelers to wild destinations around North America. His passions center on connecting people to place and providing experiences to more deeply engage people in the natural world. He received his Masters degree from University of Vermont's Field Naturalist Program.

Andlea Brett (who also goes by the name Andrea) is a Native Abenaki Vermonter. She has worked as a social worker for more than 20 years, has served on the Vermont Commission for Native American Affairs, and currently chairs the Governor's Racial Equity Advisory Panel. She is also a coleader of the University of Vermont Medical Center's BIPOC Employee Resource Group and participates in multiple governmental social equity and Indigenous rights initiatives.

Prucia Buscell is a freelance writer and editor. She was a newspaper reporter for more than 20 years and won awards for investigative and public service reporting and womens' interests writing. She has coauthored books and book chapters and written extensively about positive deviance and behavior change in healthcare, education, and other fields.

Tom Butler, a conservationist and writer, is the author, editor, or coeditor of more than a dozen books including *Wildlands Philanthropy*, *Plundering Appalachia*, *Protecting the Wild*, and *On Beauty: Douglas R. Tompkins—Aesthetics and Activism*. He formerly edited *Wild Earth* journal, was vice president for conservation advocacy at Tompkins Conservation, and was a past board president of Northeast Wilderness Trust. He currently serves as the Trust's senior fellow.

Kevin Cross is a freelance graphic designer, editor, and creative consultant, and was art director of *Wild Earth* journal. For 25 years he has sought to bring good design and strategic communication to the work of protecting and rewilding nature. He savors quiet travel by foot and canoe and delights in encounters with everything from moths to mountains.

Alicia Daniel's love of nature has taken her across the continent from tracking Alaskan black bears to surveying Mexican free-tailed bats in Texas caves to seeing an *arribada* of olive ridley sea turtles on a moonlit beach in Costa Rica. Alicia teaches in the University of Vermont Field Naturalist Program, founded the Vermont Master Naturalist Program, and works for the City of Burlington managing forested parks for wildlife and plant diversity.

Eva Duedahl is a Ph.D. Research Fellow at the Inland Norway University of Applied Sciences, Faculty of Business and Social Sciences. Eva's teaching and research include and combine tourism and leisure, sustainable development, co-design methodologies, and innovation.

Annie Faulkner is a land conservationist and an emeritus board member of Northeast Wilderness Trust. She grew up exploring the wilds of New England and the American West and earned degrees from Middlebury College, the University of Washington, and Boston University. Annie advocates for wild forests, climate security, and reproductive rights. She lives in Keene, New Hampshire, with husband Bob King, kids Nina and Ben, and two fuzzy friends.

Eric Hagen grew up in Wisconsin and graduated with a master's degree from the University of Vermont's Field Naturalist Program in 2020. Eric has a passion for understanding how ecosystems grow and change, and for learning how people fit in with the rest of nature. Eric now lives in British Columbia where he is building new friendships with unfamiliar natural communities and pursuing a career in habitat restoration and conservation planning.

Curt Lindberg's interest in nature was stimulated by his recent move to the Green Mountain State. Learning from talented Vermont naturalists and E. O. Wilson's writings spurred him to help found the Vermont Alliance for Half-Earth and join the Waitsfield Conservation Commission. His professional career has been devoted to helping people use complexity science concepts to improve the well-being of people, organizations, and the natural environment.

Maddie Lindberg is an eleven-year-old who loves nature. She, like E. O. Wilson, loves little things. Maddie lives in Montpelier, Vermont, with her family, mouse "milky way asteroid," and dog Rosie. She attends Main Street Middle School. Maddie is the granddaughter of Curt Lindberg, her "papa."

Susan Morse, the founder and science director of Keeping Track, is highly regarded as an expert in natural history and one of North America's top wildlife trackers and photographers. Since 1977, she has been monitoring wildlife, with an emphasis on documenting the presence and habitat requirements of bobcat, black bear, Canada lynx, and cougar. Sue has earned awards from numerous organizations for advancing our understanding of the natural world.

Joe Roman is a conservation biologist and marine ecologist. The author of *Listed: Dispatches from America's Endangered Species Act* and *Whale*, he has contributed to *The New York Times*, *Slate*, and *New Scientist*. He is "editor 'n' chef" of eattheinvaders.org, a website dedicated to fighting invasive species one bite at a time, and a fellow and writer in residence at the Gund Institute for Environment, University of Vermont.

George Schenk is the founder and CEO of American Flatbread at Lareau Farm and Forest. He is a past board member of the Vermont Foodbank and Vermont Business for Social Responsibility, and currently serves on the Warren Conservation Commission. He established the Lareau Interpretive Garden, a 1½-acre market garden managed for nutrient density, biodiversity, and public education, along with Community Garden plots and a Food Shelf Victory Garden.

Steven Shepard is a professional author, photographer, audio producer, and educator. He is the creator and host of the Natural Curiosity Project, a podcast devoted to the discovery of the joy and wonder of the natural world and based on the idea that curiosity leads to discovery, discovery leads to knowledge, knowledge leads to insight, and insight leads to understanding.

Arvind Singhal is the Samuel Shirley and Edna Holt Marston Endowed Professor of Communication at The University of Texas at El Paso and appointed Professor 2, Inland Norway University of Applied Sciences, Faculty of Business and Social Sciences. His teaching and research interests include diffusion of innovations, positive deviance, complexity science, and entertainment-education.

Doug Tallamy is a professor in the Department of Entomology and Wildlife Ecology at the University of Delaware, where he has authored 105 research publications and has taught insect-related courses for 40 years. He has written three books: *Bringing Nature Home*, *Nature's Best Hope*, and *The Nature of Oaks*. Doug has won several awards for his work in conservation.

Elizabeth Thompson grew up in eastern Massachusetts, where she wandered in the shrinking woods and wetlands, following old stone walls, identifying plants, and looking for flowers and frogs. These experiences fed a deep passion for nature and its conservation. Liz is a coauthor of *Wetland, Woodland, Wildland: A Guide to the Natural Communities of Vermont*.

PHOTO AND ILLUSTRATION CREDITS

Sean Beckett **covers, iii, 1, 2, 11, 20, 40, 55, 61, 79, 84, 111, 114, 171, 172**
Susan C. Morse **xvii, xviii, 48, 153, 155 (upper), 156 (lower), 159 (lower), 175, 190**
Courtesy of the E. O. Wilson Biodiversity Foundation **5**
Antonio Vizcaíno **9, 16, 100**
Tom Butler **23**
Brendan Wiltse **26, 56**
Matias Rebak/Rewilding Argentina **29**
Elizabeth Thompson **32, 39, 87**
Shelby Perry/Northeast Wildernesss Trust **34**
Harvard Forest **36–37**
Libby Davidson **45 (upper), 98**
Ken Sturm/USFWS **45 (lower)**
Zack Porter/Northeast Wildernesss Trust **46**
Courtesy of Annie Faulkner **50, 62**
Steven Shepard **66, 105, 108, 192**
Courtesy of Sandra Fary **68, 75, 80**
Joshua Brown **71**
KhteWisconsin (public domain) **82**
Lorna Dielentheis **88–89**
Lyn Baldwin **94–95**
Eric Hagen **all photos 117–146, 149 (upper and lower), 150, 155 (lower), 156 (upper and center), 159 (upper and center), 160, 163, 165 (upper), 166 (upper and lower), all photos 177–184**
University of Vermont **149 (center)**
Charlie Hohn **165 (lower), 166 (center), 168**
Curt Lindberg **197**
Flower Garden Banks National Marine Sanctuary **200**